T0215231

# Pro Couchbase Development

## A NoSQL Platform for the Enterprise

Deepak Vohra

Apress®

## Pro Couchbase Development

ISBN-13 (pbk): 978-1-4842-1435-0

ISBN-13 (electronic): 978-1-4842-1434-3

Managing Director: Welmoed Spahr
Lead Editor: Steve Anglin
Technical Reviewer: Massimo Nardone
Editorial Board: Steve Anglin, Louise Corrigan, Jonathan Gennick, Robert Hutchinson, Michelle Lowman, James Markham, Susan McDermott, Matthew Moodie, Jeffrey Pepper, Douglas Pundick, Ben Renow-Clarke, Gwenan Spearing, Steve Weiss
Coordinating Editor: Mark Powers
Copy Editor: Karen Jameson
Compositor: SPi Global
Indexer: SPi Global
Artist: SPi Global

Distributed to the book trade worldwide by Springer Science+Business Media New York, 233 Spring Street, 6th Floor, New York, NY 10013. Phone 1-800-SPRINGER, fax (201) 348-4505, e-mail orders-ny@springer-sbm.com, or visit www.springeronline.com. Apress Media, LLC is a California LLC and the sole member (owner) is Springer Science + Business Media Finance Inc (SSBM Finance Inc). SSBM Finance Inc is a Delaware corporation.

For information on translations, please e-mail rights@apress.com, or visit www.apress.com.

Apress and friends of ED books may be purchased in bulk for academic, corporate, or promotional use. eBook versions and licenses are also available for most titles. For more information, reference our Special Bulk Sales–eBook Licensing web page at www.apress.com/bulk-sales.

Any source code or other supplementary materials referenced by the author in this text is available to readers at www.apress.com/9781484214350. For detailed information about how to locate your book's source code, go to www.apress.com/source-code/. Readers can also access source code at SpringerLink in the Supplementary Material section for each chapter.

# Contents at a Glance

# Contents

# About the Author

**Deepak Vohra** is a consultant and a principal member of the NuBean.com software company. Deepak is a Sun-certified Java programmer and Web component developer. He has worked in the fields of XML, Java programming, and Java EE for over ten years. Deepak is the coauthor of *Pro XML Development with Java Technology* (Apress, 2006). Deepak is also the author of the *JDBC 4.0* and *Oracle JDeveloper for J2EE Development, Processing XML Documents with Oracle JDeveloper 11g, EJB 3.0 Database Persistence with Oracle Fusion Middleware 11g,* and *Java EE Development in Eclipse IDE* (Packt Publishing). He also served as the technical reviewer on *WebLogic: The Definitive Guide* (O'Reilly Media, 2004) and *Ruby Programming for the Absolute Beginner* (Cengage Learning PTR, 2007).

# About the Technical Reviewer

**Massimo Nardone** holds a Master of Science degree in Computing Science from the University of Salerno, Italy. He worked as a PCI QSA and Senior Lead IT Security/Cloud/SCADA Architect for many years and currently works as Security, Cloud and SCADA Lead IT Architect for Hewlett Packard Finland. He has more than twenty years of work experience in IT including Security, SCADA, Cloud Computing, IT Infrastructure, Mobile, Security, and WWW technology areas for both national and international projects. Massimo has worked as a Project Manager, Cloud/SCADA Lead IT Architect, Software Engineer, Research Engineer, Chief Security Architect, and Software Specialist. He worked as visiting a lecturer and supervisor for exercises at the Networking Laboratory of the Helsinki University of Technology (Aalto University). Massimo has been programming and teaching how to program with Perl, PHP, Java, VB, Python, C/C++, and MySQL for more than twenty years. He holds four international patents (PKI, SIP, SAML, and Proxy areas).

Massimo is the author of *Pro Android Games* (Apress, 2015).

# CHAPTER 1

■ ■ ■

# Why NoSQL?

NoSQL databases refer to the group of databases that are not based on the relational database model. Relational databases such as Oracle database, MySQL database, and DB2 database store data in tables, which have relations between them and make use of SQL (Structured Query Language) to access and query the tables. NoSQL databases, in contrast, make use of a storage and query mechanism that is predominantly based on a non-relational, non-SQL data model.

The data storage model used by NoSQL databases is not some fixed data model, but the common feature among the NoSQL databases is that the relational and tabular database model of SQL-based databases is not used. Most NoSQL databases make use of no SQL at all, but NoSQL does not imply that absolutely no SQL is used, because of which NoSQL is also termed as "not only SQL." Some examples of NoSQL databases are discussed in Table 1-1.

*Table 1-1.*  *NoSQL Databases*

| NoSQL Database | Database Type | Data Model | Support for SQL-like query language |
|---|---|---|---|
| Couchbase Server | Document | Key-Value pairs in which the value is a JSON (JavaScript Object Notation) document. | Supports N1QL, which is an SQL-like query language. |
| Apache Cassandra | Columnar | Key-Value pairs stored in a column family (table). | Cassandra Query Language (CQL) is an SQL-like query language. |
| MongoDB | Document | Key-Value pairs in which the value is a Binary JSON (BSON) document. | MongoDB query language is an SQL-like query language. |
| Oracle NoSQL Database | Key-Value | Key-Value pairs. The value is a byte array with no fixed data structure. The value could be simple fixed string format or a complex data structure such as a JSON document. | SQL query support from an Oracle database External Table. |

This chapter covers the following topics.

- What is JSON?

- What is wrong with SQL?

- Advantages of NoSQL Databases

- What has Big Data got to do with NoSQL?

- NoSQL is not without Drawbacks

- Why Couchbase Server?

- Who Uses Couchbase Server and for what?

# What Is JSON?

As mentioned in Table 1-1, the Couchbase Server data model is based on key-value pairs in which the value is a JSON (JavaScript Object Notation) document. JSON is a data-interchange format, which is easy to read and write and also easy to parse and generate by a machine. The JSON text format is a language format that is language independent but makes use of conventions familiar to commonly used languages such as Java, C, and JavaScript.

Essentially a JSON document is an object, a collection of name/value pairs enclosed in curly braces {}. Each name in the collection is followed by ':' and each subsequent name/value pair is separated from the preceding by a ','. An example of a JSON document is as follows in which attributes of a catalog are specified as name/value pairs.

```
{
"journal":"Oracle Magazine",
"publisher":"Oracle Publishing",
"edition": "January February 2013"
}
```

The name in name/value pairs must be enclosed in double quotes "". The value must also be enclosed in "" if a string includes at least a single character. The value may have one of the types discussed in Table 1-2.

*Table 1-2.* *JSON Data Types*

| Type | Description | Example |
|------|-------------|---------|
| string | A string literal. A string literal must be enclosed in "". | ```{ "c1":"v1", "c2":"v2" }```<br><br>The string may consist of any Unicode character except " and \. Each value in the following JSON document is not valid.<br><br>```{ "c1":""", "c2":"\" }```<br><br>The " and \ may be included in a string literal by preceding them with a \.<br>The following JSON document is valid.<br><br>```{ "c1":"\"", "c2":"\\" }``` |
| number | A number may be positive or negative, integer or decimal. | ```{ "c1": 1, "c2": -2.5, "c3":0 }``` |
| array | An array is a list of values enclosed in [ ]. | ```{ "c1":[1,2,3,4,5,"v1","v2"], "c2":[-1,2.5,"v1",0] }``` |
| true false | The value may be a Boolean true or false. | ```{ "c1":true, "c2":false }``` |
| null | The value may be null. | ```{ "c1":null, "c2":null }``` |
| object | The value may be another JSON object. | ```{ "c1":{"a1":"v1", "a2":"v2", "a3":[1,2,3]}, "c2":{"a1":1, "A2":null, "a3":true}, "c3":{} }``` |

The JSON document model is most suitable for storing unstructured data, as the JSON objects can be added in a hierarchical structure creating complex JSON documents. For example, the following JSON document is a valid JSON document consisting of hierarchies of JSON objects.

```
{
  "c1": "v1",
  "c2": {
    "c21":[1,2,3],
    "c22":
    {
      "c221":"v221",
      "c222":
      {
        "c2221":"v2221"
      },
      "c223":
      {
        "c2231":"v2231"
      }
    }
  }
}
```

# What Is Wrong with SQL?

NoSQL databases were developed as a solution to the following requirements of applications:

- Increase in the volume of data stored about users and objects, also termed as big data.

- Rate at which big data influx is increasing.

- Increase in the frequency at which the data is accessed.

- Fluctuations in data usage.

- Increased processing and performance required to handle big data.

- Ultra-high availability.

- The type of data is unstructured or semi-structured.

SQL-based relational databases were not designed to handle the scalability, agility, and performance requirements of modern applications using real-time access and processing big data. While most RDBMS databases provide scalability and high availability as features, Couchbase Server provides higher levels of scalability and high availability. For example, while most RDBMS databases provide replication within a datacenter, Couchbase Server provides Cross Datacenter Replication (XDCR), which is replication to multiple, geographically distributed datacenters. XDCR is discussed in more detail in a later section. Couchbase Server also provides rack awareness, which traditional RDBMS databases don't.

Big data is growing exponentially. Concurrent users have grown from a few hundred or thousand to several million for applications running on the Web. It is not just that once big data has been stored new data is not added. It is not just that once a web application is being accessed by millions of users it shall continue to be accessed by as many users for a predictable period of time. The number of users could drop to a few thousand within a day or a few days. Relational database is based on a single server architecture. A single database is a single point of failure (SPOF). For a highly available database, data must be distributed across a cluster of servers instead of relying on a single database. NoSQL databases provide the distributed, scalable architecture required for big data. "Distributed" implies that data in a NoSQL database is distributed across a cluster of servers. If one server becomes unavailable another server is used. The "distributed" feature is a provision and not a requirement for a NoSQL database. A small scale NoSQL database may consist of only one server.

The fixed schema data model used by relational databases makes it necessary to break data into small sets and store them in separate tables using table schemas. The process of decomposing large tables into smaller tables with relationships between tables is called database normalization. Normalized databases require table joins and complex queries to aggregate the required data. In contrast, the JSON document data model provided by NoSQL databases such as Couchbase provide a denormalized database. Each JSON document is complete unto itself and does not have any external references to other JSON documents. Self-contained JSON documents are easier to store, transfer, and query.

# Advantages of NoSQL Databases

In this section I'll cover the advantages of NoSQL databases.

## Scalability

NoSQL databases are easily scalable, which provides an elastic data model. Why is scalability important? Suppose you are running a database with a fixed capacity and the web site traffic fluctuates, sometimes rising much in excess of the capacity, sometimes falling below the capacity. A fixed capacity database won't be able to serve the requests of the load in excess of the capacity, and if the load is less than the capacity the capacity is not being utilized fully. Scalability is the ability to scale the capacity to the workload. Two kinds of scalability options are available: *horizontal scalability* and *vertical scalability*. With horizontal scalability or scaling-out, new servers/machines are added to the database cluster. With vertical scalability or scaling-up, the capacity of the same server or machine is increased. Vertical scalability has several limitations.

- Requires the database to be shut down so that additional capacity may be added, which incurs a downtime.

- A single server has an upper limit.

- A single server is a single point of failure. If the single server fails, the database becomes unavailable.

While relational databases support vertical scalability, NoSQL databases support horizontal scalability. Horizontal scalability does not have the limitations that vertical scalability does. Additional server nodes may be added to a Couchbase cluster without a dependency on the other nodes in the cluster. The capacity of the NoSQL database scales linearly, which implies that if you add four additional servers to a single server, the total capacity becomes five times the original, not a fraction multiple of the original due to performance loss. The NoSQL cluster does not have to be shut down to add new servers. Ease of scalability is provided by the shared-nothing architecture of NoSQL databases. The monolithic architecture provided by traditional SQL databases is not suitable for the flexible requirements of storing and processing big data. Traditional databases support scale-up architecture (vertical scaling) in which additional resources may be added to a single machine. In contrast, NoSQL databases provide a scale-out (horizontal scaling), nothing shared architecture, in which additional machines may be added to the cluster. In a shared-nothing architecture, the different nodes in a cluster do not share any resources, and all data is distributed (partitioned) evenly (load balancing) across the cluster by a process called sharding.

## Ultra-High Availability

Why is high availability important? Because interactive real-time applications serving several users need to be available all the time. An application cannot be taken offline for maintenance, software, or hardware upgrade or capacity increase. NoSQL databases are designed to minimize downtime, though different NoSQL databases provide different levels of support for online maintenance and upgrades. Couchbase Server supports online maintenance, software and hardware upgrades, and scaling-out. As mentioned earlier, Couchbase Server provides ultra-high availability.

## Commodity Hardware

NoSQL databases are designed to be installed on commodity hardware, instead of high-end hardware. Commodity hardware is easier to scale-out: simply add another machine and the new machine added does not even have to be of similar specification and configuration as the machine/s in the NoSQL database cluster.

## Flexible Schema or No Schema

While the relational databases store data in the fixed tabular format for which the schema must be defined before adding data, the NoSQL databases do not require a schema to be defined or provide a flexible dynamic schema. Some NoSQL databases such as Oracle NoSQL database and Apache Cassandra have a provision for a flexible schema definition, still others such as Couchbase are schema-less in that the schema is not defined at all. Any valid JSON document may be stored in a Couchbase Server. One document may be different from another and the same document may be modified without adhering to a fixed schema definition. The support for flexible schemas or no schemas makes NoSQL databases suitable for structured, semi-structured, and unstructured data. In an agile development setting the schema definition for data stored in a database may need to change, which makes NoSQL databases suitable for such an environment. Dissimilar data may be stored together. For example, in the following JSON document the c21 name has an array of dissimilar data types as value.

```
{
  "c1": "v1",
    "c21":[1,"c213", 2.5, null, true]
}
```

In contrast, a value in a relation database column must be of the schema definition type such as a string, an integer, or a Boolean. Flexible schemas make development faster, code integration uninterrupted by modifications to the schema, and database administration almost redundant.

## Big Data

NoSQL databases are designed for big data. Big data is in the order of tens or even hundreds of PetaByte (PB). For example, eBay, which makes use of Couchbase stores 5.3 PB on a 532 node cluster. TuneWiki uses Couchbase to store more than one billion documents. Big data is usually associated with a large number of users and a large number of transactions. Viber, a messaging and VoIP services company handles billions of messages a month and thousands of ops per second with Couchbase for its big data requirements.

# Object-Oriented Programming

The key-value data model provided by NoSQL databases supports object-oriented programming, which is both easy to use and flexible. Most NoSQL databases are supported by APIs in object-oriented programming languages such as Java, PHP, and Ruby. All client APIs support simple put and get operations to add and get data.

# Performance

Why is performance important? Because interactive real-time applications require low latency for read and write operations for all types and sizes of workloads. Applications need to serve millions of users concurrently at different workloads. The shared-nothing architecture of NoSQL databases provides low latency, high availability, reduced susceptibility to failure of critical sections, and reduced bandwidth requirement. The performance in a NoSQL database cluster does not degrade with the addition of new nodes.

# Failure Handling

NoSQL databases typically handle server failure automatically to failover to another server. Why is auto-failover important? Because if one of the nodes in a cluster were to fail and if the node was handling a workload, the application would fail and become unavailable. NoSQL databases typically consist of a cluster of servers and are designed with the failure of some nodes as expected and unavoidable. With a large number of nodes in a cluster the database does not have a single point of failure, and failure of a single node is handled transparently with the load of the failed server being transferred to another server. Couchbase keeps replicas (up to three) of each document across the different nodes in the cluster with a document on a server being either in active mode or as an inactive replica. The map of the different document replicas on the different servers in the cluster is the cluster topology. The client is aware of the cluster topology. When a server fails, one of the inactive replica is promoted to active state, and the cluster topology is updated, without incurring any downtime as is discussed in a later section.

# Less Administration

NoSQL databases are easier to install and administer without the need for specialized DBAs. A developer is able to handle the administration of a NoSQL database, but a specialized NoSQL DBA should still be used. Schemas are flexible and do not need to be modified periodically. Failure detection and failover is automatic without requiring user intervention. Data distribution in the cluster is automatic using auto-sharding. Data replication to the nodes in a cluster is also automatic. When a new server node is added to a cluster, data gets distributed and replicated to a new node as required automatically.

# Asynchronous Replication with Auto-Failover

Most NoSQL databases such as Couchbase provide *asynchronous* replication across the nodes of a cluster. Replication is making a copy of data and storing the data in a different node in the cluster. Couchbase stores up to three replicas. The replication is illustrated in Figure 1-1 in which a JSON document is replicated to three nodes in a Couchbase cluster. On each node the document is available either in Active state or as a passive Replica. If the Active document on a node becomes unavailable due to server failure or some reason such as power failure, a replica of the document on another server is promoted to Active state. The promotion from Replica passive state to Active Sate is transparent to the client without any downtime or very less downtime.

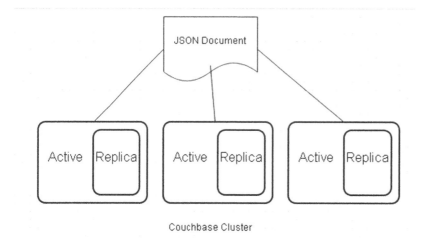

Couchbase Cluster

**Figure 1-1.** *Replication on a Couchbase Cluster*

Replication within a cluster provides durability, reliability, and high availability in the eventuality of a single node failure. The terms durability, reliability, and high availability seem similar but have different connotations. Durability is a measure of the time for which the data is not lost and is in a persistent state. Reliability is a measure of the operational efficiency of the database. Common measures of reliability are Mean Time Between Failure (MTBF=total time in service/number of failures during the same time) and Failure rate (number of failures/total time in service). High availability is the measure of time for which the database is available; Available time/(Available time+Not Available time).

Couchbase also supports Cross Datacenter Replication (XDCR), which is replication of data from one data center to another. In addition to failure recovery, XDCR provides data locality because, with the same data replicated across multiple data centers, it is more likely to find a cluster/node closer to a client. I cover XDCR in the section "Cross Data Center Replication."

"Asynchronous" implies that a server does not wait for the replication to complete before sending an ACK (acknowledgment) to the client. The difference between *synchronous* and *asynchronous* mode is explained next. In synchronous mode a data is replicated in the following sequence and illustrated in Figure 1-2.

1. Client sends a new data record to Server1.

2. The data record is stored in NoSQL database on Server1.

3. The data record is propagated to Server2 for replication.

4. The data record is stored in NoSQL database on Server2.

5. The Server2 sends ACK to Server1 that the data record has been replicated.

6. The Server1 sends ACK to Client that the data record has been replicated.

**Synchronous Mode**

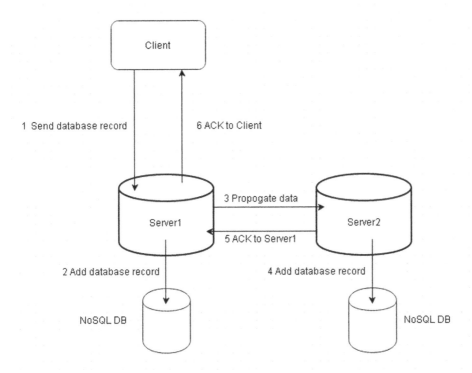

***Figure 1-2.*** *Data Replication in Synchronous Mode*

In asynchronous mode, a data is replicated in the following sequence and illustrated in Figure 1-3.

1. Client sends a new data record to Server1.

2. The data record is stored in NoSQL database on Server1.

3. The data record is propagated to Server2 for replication.

4. The Server1 sends ACK to Client that the data record has been replicated.

5. The data record is stored in NoSQL database on Server2.

6. The Server2 sends ACK to Server1 that the data record has been replicated.

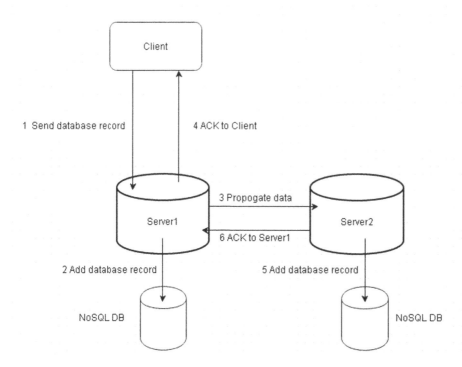

**Figure 1-3.** *Data Replicationin Asynchronous Mode*

Asynchronous mode prevents the latency associated with waiting for a response from the servers to which data is propagated for replication. But, the servers in the cluster could be an inconsistent state while data is being replicated. The client, however, gets an ACK for data replication before the consistent state is stored. Data in asynchronous mode is eventually consistent.

## Caching for Read and Write Performance

Most NoSQL databases, including Couchbase Server, provide integrated object-level caching to improve read and write performance. With caching, applications are able to read and write data with a latency of less than a millisecond. Caching improves read performance more than it improves write performance.

## Cloud Enabled

Cloud computing has made unprecedented capacity and flexibility in choice of infrastructure available. Cloud service providers such as Amazon Web Services (AWS) provide fully managed NoSQL database services and also the option to develop custom NoSQL database services. AWS has partnered with Couchbase to provide support and training to those running Couchbase Server on Amazon EC2 and Amazon EBS.

# What Has Big Data Got to Do with NoSQL?

Though NoSQL databases may be used for storing small quantities of data, NoSQL databases were motivated by big data and the dynamic requirements of big data storage and processing. Couchbase Server is designed for big data with features such as scalability, intra cluster, and cross datacenter replication. In some of the examples in the book we shall use small quantities of data to demonstrate features and client APIs. The quantity of data stored or fetched may be scaled as required in a big data application. The same application that is used to stored ten lines of data in Couchbase may be modified to store a million lines of data. The same application that is used to migrate five rows of data from Apache Cassandra to Couchbase Server may be used to migrate a million rows of data. The performance of Couchbase Server does not deteriorate with increase in data processed.

# NoSQL Is Not without Drawbacks

While much has been discussed about their merits, NoSQL databases are not without drawbacks. Some of the aspects in which NoSQL databases have limitations are as follows.

## BASE, Not ACID

NoSQL databases do not provide the ACID (Atomicity, Consistency, Isolation, and Durability) properties in transactions that relational databases do.

- Atomicity ensures that either all task/s within a transaction are performed or none are performed.

- Consistency ensures that the database is always in a consistent state without any partially completed transactions.

- Isolation implies that transactions are isolated and do not have access to the data of other transactions until the transactions have completed. Isolation provides consistency and performance.

- A transaction is durable when it has completed.

NoSQL database provide BASE (Basically Available, Soft state, and Eventually consistent) transactional properties.

- Basically Available implies that a NoSQL database returns a response to every request though the response could be a failure to provide the requested data, or the requested data could be returned in an inconsistent state.

- Soft state implies that the state of the system could be in transition during which time the state is not consistent.

- Eventually consistent implies that when the database stops receiving input, eventually the state of a NoSQL database becomes consistent when the data has replicated to the different nodes in the cluster as required. But, while a NoSQL database is receiving input, the database does not wait for its state to become consistent before receiving more data.

## Still New to the Field

The NoSQL databases are still new to the field of databases and not as functionally stable and reliable as the established relational databases.

## Vendor Support Is Lacking

Most NoSQL databases such as MongoDB and Apache Cassandra are open source projects and lack the official support provided by established databases such as Oracle database or IBM DB2 database. Couchbase Server is also an open source project. Couchbase, however does provide subscription-based support for its Enterprise Edition server.

# Why Couchbase Server?

Couchbase Server is a high-performance, distributed, NoSQL database. Couchbase Server provides several benefits additional or similar to those provided by some of the other leading NoSQL databases.

## Flexible Schema JSON Documents

Interactive, real-time applications, processing unstructured data required to support a varying data model as the unstructured data does not conform to any fixed schema. Not all NoSQL databases are based on the JSON data model. In Couchbase Server, data is stored as JSON documents with each document assigned an Id. The JSON data storage and exchange format is a schema-less data model as discussed earlier and stores hierarchies of name/value pairs. The JSON document structure is not fixed and may vary from document to document and may be modified in the same document. The only requirement is that the document is a valid JSON document. A flexible schema data model does not require an administrator's intervention to modify schema, which could lead to downtime.

## Scalability

While all NoSQL databases are scalable Couchbase's scalability feature has the following advantages.

- Adding and removing nodes is a one-click solution without incurring downtime. All nodes are the same type, which precludes the requirement to configure different types of nodes.

- Auto-sharding, which is discussed in more detail in the next subsection, provides automatic load balancing across the cluster with no hot spots on overloaded servers.

- The Cross Data Center Replication feature is unique to Couchbase and makes Couchbase scalable across geographies.

## Auto-Sharding Cluster Technology

When a new server is added or removed from a Couchbase cluster, data is automatically redistributed to the nodes in the cluster and rebalanced without downtime in serving client requests. The process of evenly distributing data across the cluster automatically is called auto-sharding. If more RAM and I/O capacity is required, simply add a server. Data is available continuously while being balanced evenly among the cluster nodes. Client requests are routed to a server closest to the client making use of data locality. Data locality improves response time and reduces network traffic as data is being served from a server that is close to the client.

# High Performance from High Throughput and Low Latency

Latency may be defined in different forms but all imply a delay: for example, the delay in receiving requested data or a delay in data transfer to another server for replication. Throughput is defined as the rate of data transfer over a network.

Couchbase is designed for the flexible data management requirements of interactive web applications providing high throughput and low latency. While most NoSQL databases provide a fast response, Couchbase's sub-millisecond latency is consistent across read and write operations and consistent across varying workloads. The latency of some of the other NoSQL databases such as MongoDB and Apache Cassandra increases as the number of ops/sec increases, but Couchbase's latency stays low even at high workloads. While most NoSQL databases provide a high throughput, Couchbase's high throughput is consistent across a mix of read and write operations. Throughput scales linearly with additional nodes. In a performance benchmark (`http://www.slideshare.net/renatko/couchbase-performance-benchmarking`) comparing Apache Cassandra, MongoDB, and Couchbase, Couchbase showed the lowest latencies and highest throughput. One of the reasons Viber cited for choosing Couchbase was that "Couchbase was able to provide several times more throughput using less than half the number of nodes."

Couchbase provides built-in *memcache*-based caching technology. What is memcache? Memcache is a cache in the memory (RAM) to store temporarily (also called to cache) frequently used data. Memcache is used to optimize disk I/O; if data is made available from the RAM the disk does not have to be accessed. Memcache is also used to optimize CPU; results of CPU intensive computations are stored in the cache to avoid recomputation. What is "frequently used data" is determined by the server based on the number and frequency of requests for the data. The RAM not being used for other purposes is used as memcache, and memcache is temporary as the RAM may be reclaimed for other use if required. Couchbase Server coordinates with the disk to keep sufficient RAM to serve incoming requests with low latency for high performance. When the frequently used information is re-requested it is served from the memcache instead of fetching from the database. Memcache improves response time, which results in reduced latency and high throughput. With sub-millisecond read and write performance, Couchbase Server is capable of hundreds of thousands of ops per second per server node. Couchbase Server persists data from RAM to disk asynchronously while keeping a set of data for client access in the object-level cache in RAM. An append-only storage tier appends data contiguously to the end of a file, improving performance. Updates are first committed to RAM and subsequently to disk using per-document commit. A cache miss is defined as a direct access of a database disk when the cache does not provide the required data. Orbitz mentioned caching mechanism as the main reason for choosing Couchbase.

# Cluster High Availability

Couchbase cluster stays highly available without downtime. While most NoSQL databases provide high availability, Couchbase has the following advantages over the others.

- Cross Data Center Replication, which is discussed in detail in the next section, provides high availability even in the eventuality of a whole data center failing.

- Software upgrades are done online, without shutting down the Couchbase Server.

- Hardware upgrades are done online.

- Maintenance operations such as compaction are done online.

# Cross Data Center Replication

Replication of data stores multiple copies of data on different nodes in a cluster for durability and high availability. Durability implies that if one copy of the data is lost due to machine failure or some other reason such as power failure, another copy of the data is still available. High Availability implies that the database does not have downtime due to the failure of a single node in the cluster as a copy of the data from another node is fetched. In additional to replication within a cluster (intra cluster replication), Couchbase 2.0 added a feature called *Cross Data Center Replication* (XDCR) in which data is replicated across data centers to cluster/s in another data center, which could be at a geographically remote location. XDCR provides data locality in addition to the benefits discussed previously in this section. Data locality is the closeness of data to a client. If each client is able to access a node that is close to the client, data is not required to be transmitted across the network. If data is available at a data center close to a client, data is fetched from the data center instead of fetching over the network from a distant data center. Transmitting data across the network incurs delay (latency) and increased bandwidth requirement. Data locality improves response time. Cross Datacenter Replication is illustrated in Figure 1-4 in which a JSON Document A is replicated using intra cluster replication on Datacenter1, and JSON Document B is replicated using intra cluster replication on Datacenter2. JSON document is also replicated using XDCR on Datacenter2 and JSON Document B is replicated using XDCR on Datacenter1. The number of replicas may vary based on requirement.

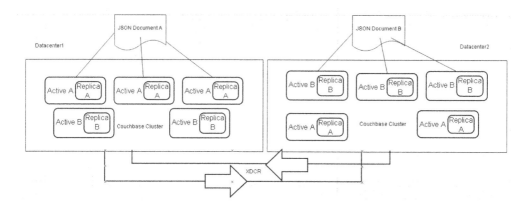

***Figure 1-4.*** *Cross Data Center Replication (XDCR)*

XDCR replicates data unidirectionally or bidirectionally between data centers. With bidirectional replication, data may be added in either data center and read from another data center.

## Data Locality

Data locality is the closeness of a Couchbase Server to its client. Cross Data Center Replication makes it feasible to replicate data across geographies. A client is served from a data center that is closest to the client, thereby reducing the network latency.

## Rack Awareness

Couchbase Servers in a cluster are stored across several racks and each rack has its own power supply and switches. Failure of a single rack makes data stored on the rack susceptible to loss. To prevent loss of all copies of a data and provide high availability, Couchbase Server 2.5 Enterprise Edition introduced Rack Awareness. Using Couchbase, Rack Awareness replicas of a document are placed on nodes across different racks so that failure of a single rack does not cause all replicas of the document to be lost or become unavailable, even temporarily.

## Multiple Readers and Writers

As of Couchbase Server 2.1, multiple readers and writers are supported to persist/access data to/from a disk to fully utilize the increase in disk speeds to provide high read and write efficiency. With a single thread read and write, the data in the cache is less as compared to data on the disk resulting in cache misses, which results in increased response time and increased latency. With multiple threads accessing the same disk, more data may be fetched into the cache to improve efficiency of read and write to improve the response time and reduce the latency. Multithreaded engine includes synchronization among threads to prevent multiple threads from accessing the same data concurrently.

## Support for Commonly Used Object-Oriented Languages

Couchbase Server provides client APIs for commonly used languages such as Java, PHP, Ruby, and C.

## Administration and Monitoring GUI

Couchbase provides administration and monitoring graphical user interface (GUI), which some of the other NoSQL databases such as MongoDB don't. Some third-party admin GUIs are available for MongoDB but a built-in integrated admin GUI is not provided.

# Who Uses Couchbase Server and for What?

A wide spectrum of companies from different industries use Couchbase Server. Different companies have different reasons for choosing Couchbase Server. Reasons cited by some of the companies who chose Couchbase are listed in Table 1-3.

***Table 1-3.*** *Reasons for Using Couchbase*

| Company | Reasons |
| --- | --- |
| AOL for ad targeting | AOL uses Couchbase in conjunction with Hadoop to make hundreds of user profiles and statistics available for their ad targeting platform with sub-millisecond latency. |
| DOCOMO Innovations for mobile services | Real-time data infrastructure, mobile-to-cloud-data synchronization, elastic scalability, production-ready solution with high availability. |
| OMGPOP for Draw Something | Scalability without downtime or performance degradation. |
| Orbitz for travel services | Couchbase provides no downtime. Couchbase is used store user online sessions. Couchbase provides integrated Memcache for fast response. |

*(continued)*

***Table 1-3.*** (*continued*)

| Company | Reasons |
|---|---|
| Betfair for online betting | Scalability and Replication with auto-failover are suitable for Betfair's Continuous Delivery methodology. Betfair processes more than 7 million transactions per day with each completing in less than a second. Betfair uses Couchbase to store session data across sessions and for storing user preferences for customization. Couchbase provides high performance, scalability, schema flexibility and continuous delivery. |
| AdAction for ad serving | Couchbase is used to store large quantities of consumer data for about 75 million users per month. Couchbase chosen because of its performance, uptime, high response time, low administrative overhead, scalability without performance loss, and rapid deployment. |
| Amadeus for travel services | Couchbase server was chosen because of its low (sub-millisecond) latency, elasticity to handle traffic growth, high throughput and linear scalability when adding/removing nodes. |
| Concur for business travel | As Concur processes more than a billion Couchbase operations per day Couchbase's low latency was one of the reasons for being chosen. Couchbase's cluster management made it feasible to add/remove nodes without downtime. Couchbase's seamless transition when adding/removing nodes requires no configuration management with all clients being updated automatically. A single solution for multiple tiers and languages was one of the main reasons for choosing Couchbase. |
| LinkedIn for professional social networking | With hundreds of millions of users LinkedIn chose Couchbase for its performance and scalability that can be used for logging, monitoring and analyzing the metrics of user activity. High availability caching was one of the main reasons for choosing Couchbase. |
| Nami Media for enterprise class advertising solutions | Couchbase was chosen for its fault-tolerance, data persistence and high availability. Linear scalability with no downtime and Couchbase's monitoring of the cluster to provide RAM and disk persistence statistics were some of the other reasons. |

Couchbase users include AOL, Orbitz, Cisco, LinkedIn, and Concur. Table 1-4 lists additional Internet companies and Enterprises who use Couchbase Server.

***Table 1-4.*** *Some Companies Using Couchbase*

| Internet Companies | Enterprises |
|---|---|
| AOL | LG |
| Orbitz | ADP |
| LinkedIn | Cisco |
| adscale | NTT Docomo |
| Ubisoft | Vodafone |
| Tapjoy | Skechers |
| Dotomi | NCR Corporation |
| Playdom | Comcast |
| Concur | ITT |
| Sabre | Experian |

Couchbase Server is used for a wide variety of applications ranging from advertising to VoIP services. Couchbase Server and NoSQL, in general, are used for applications storing and processing big data. Examples of types of applications using NoSQL are discussed in Table 1-5.

***Table 1-5.*** *Types of Applications Using Couchbase*

| Application | Example |
|---|---|
| User profile management distributed globally | The user profile of millions of LinkedIn, Tunewiki, and AOL users is stored in Couchbase NoSQL database. The semi-transient device data of millions of musiXmatch users is stored in Couchbase Server. |
| Session store management | The user sessions of millions of clients who log on to Orbitz, Concur, Sabre, and musiXmatch are stored in Couchbase database. |
| Content and metadata store management | Some of the challenges in content and metadata store management are: Content and metadata are unstructured. Scalability to support millions of concurrent users. High-performance interactive, customized applications. Search across the full dataset. Couchbase is suitable for the following reasons: |
| | • Elastisearch provides real-time, integrated, distributed, full-text search. |
| | • Flexible data model to provide a wide variety of data. |
| | Scalability for fluctuations in workload. High performance with low latency and high throughput. No downtime. |

(*continued*)

*Table 1-5.* (*continued*)

| Application | Example |
|---|---|
| Data aggregation | Couchbase is used to store social media data. Used for data aggregation by Sambacloud. |
| High-availability cache | Couchbase Server is used as the cache tier by Orbitz. Couchbase is suitable for providing a caching layer to replace a separate Memcache tier. A separate Memcache tier has several drawbacks such as cold cache, heavy RDBMS contention, lack of scalability, complex monitoring, and stale data access. |
| Mobile Apps | Couchbase is used to store user info and app content by Kobo and Playtika. |
| Ad Targeting | AOL makes use of Couchbase for advertising targeting in real time. Couchbase provides fast access. Some others using Couchbase for ad targeting are Dotomi, Nami Media, Xclaim, adscale, Chango, Delta Project's Ad Action. The following features of Couchbase make it suitable for ad targeting. Production proven in large-scale ad and offer targeting systems. Schema-less data model. Elastic scaling on commodity hardware or cloud computing instances. Sub-millisecond read and write latency. Hadoop support. Built-in caching. |
| Social Gaming | Couchbase stores player and game data, for example: Tapjoy, Ubisoft, Pokemon, Quepasa, Antic Entertainment, Gamegos, Meteor Games, Nexon, Playtika, Scoreloop, Shuffle Master, Sojo Studios, Tribal Crossing , Betfair, VNG, Vostu. The following features make Couchbase suitable for social gaming: Sub-millisecond response time. No downtime provides interruption-free platform. Flexible schema provides rapid game development. |
| Communications | NTT Docomo, VodafoneCisco, ITT. Couchbase's ability to provide real-time data makes it suitable for the Communications industry. |
| Business Services | Salesforce.com, ADP, Concur, Deutsche Post, LG CNS, Navteq |
| E-Commerce | Skechers, Ganz, Kobo, Skyscanner, The Knot |
| Social networking | Vimeo, mig33, Spotme, Tango, The Ladders, Tunewiki |

# Summary

This chapter introduced the NoSQL databases. We discussed the JSON data format, which is the format used in Couchbase Server. We discussed the advantages of NoSQL databases over SQL-relational databases. We introduced the Couchbase Server and why it is one of the best NoSQL databases. We also listed some of the users of Couchbase Server. We introduced NoSQL in the context of Big Data. In the next chapter, we shall discuss accessing Couchbase Server with a Java client and running CRUD operations in the database.

# CHAPTER 2

# Using the Java Client

Couchbase Server is a document object store based on the flexible JSON model. Unlike relational databases, which support only the fixed row-column model, no fixed schema is required for the Couchbase Server. Hierarchies of JSON structure can be constructed to develop a custom data model. With fast (sub-millisecond) response times, high throughput, scalability, view-based querying, and support for Map & Reduce function, Couchbase Server is one of the leading NoSQL databases, if not the leader, in NoSQL databases. In this chapter we shall create a document store in Couchbase Server using the Java Client Library for Couchbase Server. We shall use Eclipse IDE for developing a Java application to access Couchbase Server. This chapter covers the following topics.

- Setting up the Environment
- Creating a Maven Project
- Connecting to Couchbase Server
- Creating a Data Bucket
- Creating a Document
- Getting a Document
- Updating a Document
- Creating a View
- Querying a View
- Deleting a Document

## Setting Up the Environment

Download and install the following software components.

- Couchbase Server for Windows, Community or Enterprise Edition, from http://www.couchbase.com/download.
- Couchbase Java Client Library 2.1.3 as Maven dependency.
- Eclipse IDE for Java Developers (Eclipse Luna SR2) from http://www.eclipse.org/downloads/.
- JDK 1.7 or 1.8 from http://www.oracle.com/technetwork/java/javase/downloads/.

# Creating a Maven Project

In this section we shall create a Maven project in Eclipse for the Couchbase Java Client to access the Couchbase Server. We shall also add a Java class to the Maven project.

1.  Select File ➤ New ➤ Other. Select Maven ➤ Maven Project in New as shown in Figure 2-1. Click on Next.

***Figure 2-1.*** *Creating a New Maven Project*

2.  In New Maven Project window, select the Create a simple project check box and the Use default Workspace location check box as shown in Figure 2-2. Click on Next.

***Figure 2-2.*** *Selecting New Maven Project name and location*

3. As Maven project configuration, specify Group Id, Artifact Id, Version, Packaging and Name as shown in Figure 2-3. Click on Finish.

***Figure 2-3.*** *Configuring New Maven Project*

4. Next, add a Java application to the Maven project. Select File ➤ New ➤ Other as before. In New window, select Java ➤ Java Class as shown in Figure 2-4. Click on Next.

***Figure 2-4.*** *Creating a New Java Class*

5.  In New Java Class, select the Source folder as CouchbaseJava/src/main/java, specify Package as couchbase and class Name as CouchbaseJavaClient as shown in Figure 2-5. Select the check box **public static void main (String[] args)** and click on Finish.

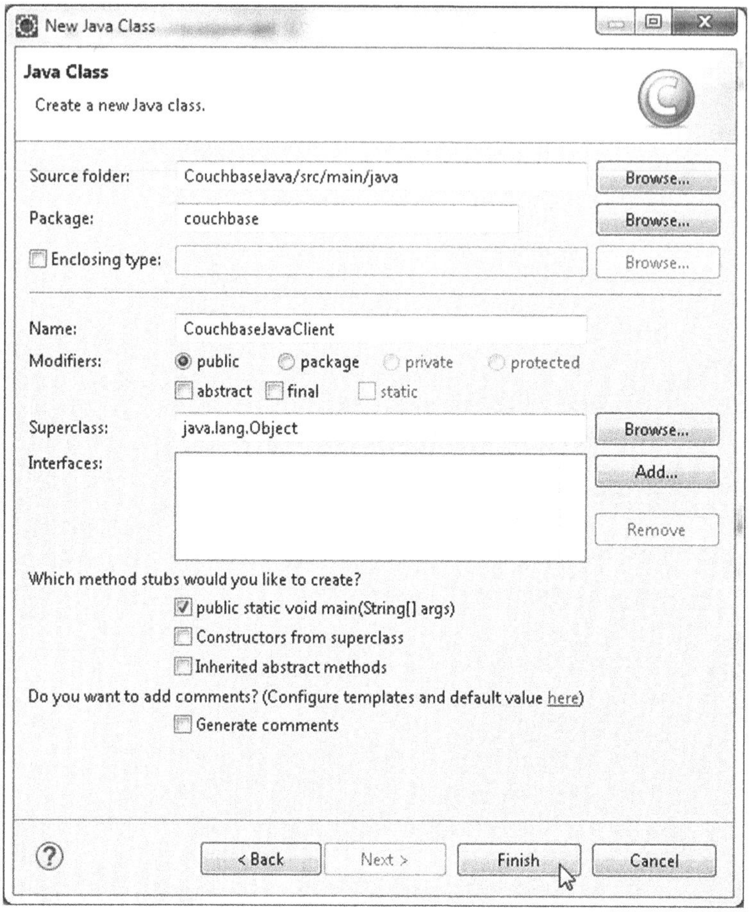

**Figure 2-5.** *Creating New Java Class*

6. A Java class gets created in Maven project as shown in the Package Explorer in Figure 2-6.

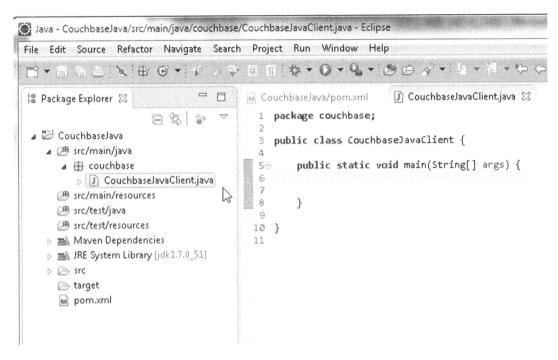

***Figure 2-6.** New Java Class CouchbaseJavaClient*

7. Next, add the Maven dependency for Couchbase Java Client to the pom.xml. Right-click on pom.xml in Package Explorer and select Open With ➤ Maven POM Editor as shown in Figure 2-7.

***Figure 2-7.** Maven POM Configuration File pom.xml*

8. The Couchbase Java Client Library 2.1.3 and dependency jar files get added to the project build path. Right-click on the CouchbaseJavaClient project node in Package Explorer and select Properties. In Properties window select the Java Build Path node and subsequently the Libraries tab. The Maven jar files are listed in the Java Build Path as shown in Figure 2-8. Click on OK.

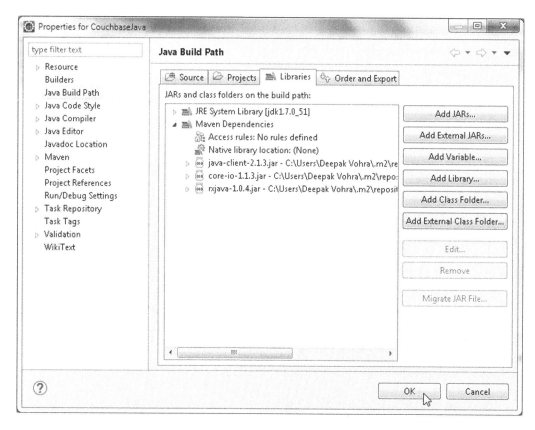

***Figure 2-8.*** *Maven Dependency Jars*

The pom.xml configuration file is listed below.

```xml
<project xmlns="http://maven.apache.org/POM/4.0.0" xmlns:xsi="http://www.w3.org/2001/
XMLSchema-instance"
    xsi:schemaLocation="http://maven.apache.org/POM/4.0.0 http://maven.apache.org/xsd/maven-
                        4.0.0.xsd">
    <modelVersion>4.0.0</modelVersion>
    <groupId>couchbase.client.java</groupId>
    <artifactId>CouchbaseJava</artifactId>
    <version>1.0.0</version>
    <name>CouchbaseJava</name>
    <dependencies>
        <dependency>
            <groupId>com.couchbase.client</groupId>
            <artifactId>java-client</artifactId>
            <version>2.1.3</version>
        </dependency>
    </dependencies>
</project>
```

In the following sections we shall connect to the Couchbase Server using the Couchbase Java Client library and add, get, update, and delete document/s. We shall also query Couchbase using a view. Add the methods listed in Table 2-1 to CouchbaseJavaClient.

**Table 2-1.** *Methods in CouchbaseJavaClient Class*

| Constructor | Description |
| --- | --- |
| createDocument() | Creates a Couchbase document |
| updateDocument() | Updates a document |
| getDocument() | Gets a document |
| removeDocument() | Deletes a document |
| queryView() | Queries a view |

Documents may be stored in the default bucket or a user created bucket. In the next section we shall discuss creating a bucket.

## Creating a Data Bucket

Couchbase Server stores data in Data buckets. The "default" data bucket is created by default when the Couchbase Server is installed. The data bucket supports items up to 20MB in size. Additional data buckets may be added as required. A new data bucket could be added for a new application or if the bucket limit is reached. A new data bucket may be created in the Couchbase Console. If the available RAM is not sufficient to create multiple buckets the RAM/Quota Usage for one of the buckets may have to be reduced. For example, reduce the RAM/Quota Usage for the default bucket to 100 MB. In the Console select the Data Buckets link and click on Create New Data Bucket button to create a new data bucket as shown in Figure 2-9.

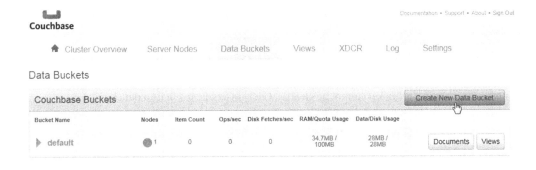

**Figure 2-9.** *Creating a New Data Bucket in Couchbase Server from the Admin Console*

In the Create Bucket pop-up specify a Bucket Name, json, for example, as shown in Figure 2-10. Select Bucket type as Couchbase, Memcached being the other. The Couchbase bucket type supports the full range of Couchbase-specific functionality and also has a higher Item size (20 MB) in contrast to 1 MB for Memcached. Memcached bucket type provides none of the advantages of Couchbase bucket type such as replication and XDCR. In fact, Memcached does not provide persistence and is just an in-memory cache

27

designed to be used alongside a relational database for frequently used data. Select the default setting for the Memory Size. For Access control specify a password. Replicas are not enabled by default. If replicas are to be created, select Replicas ➤ Enable.

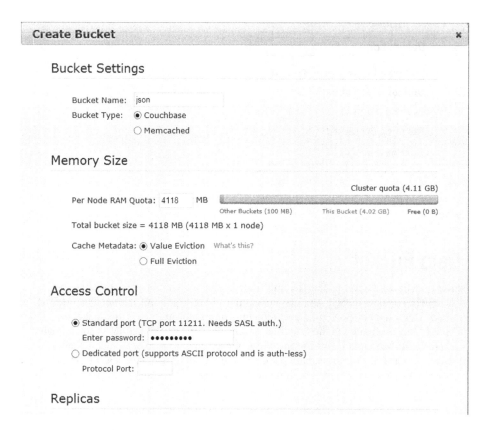

***Figure 2-10.*** *Configuring the Data Bucket*

Enable Flush as shown in Figure 2-11. The Flush option makes the data in bucket deletable; without the Flush option the data in the bucket cannot be cleared. Click on Create to create a data bucket.

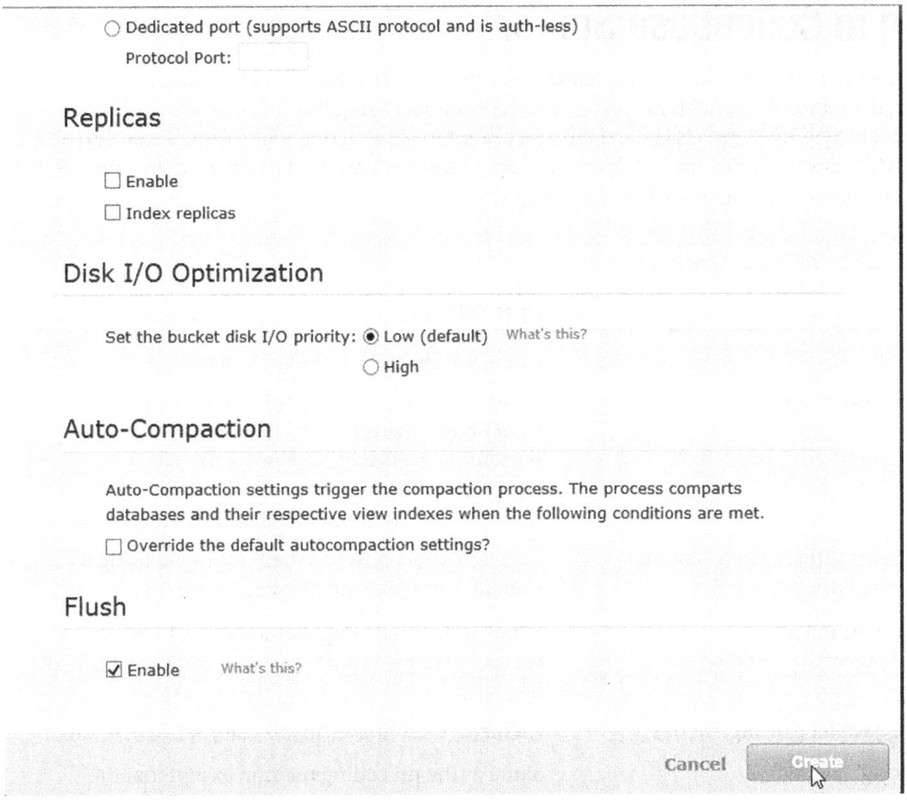

***Figure 2-11.*** *Enabling Flush and Creating the Data Bucket*

A new data bucket gets created as shown in Figure 2-12.

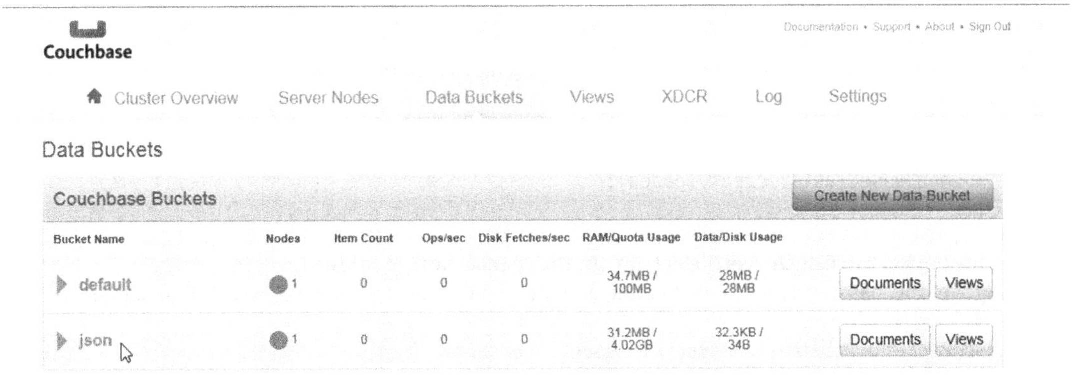

***Figure 2-12.*** *New Data Bucket*

# Connecting to Couchbase Server

In this section we use the Java client class for Couchbase Server, com.couchbase.client.java. CouchbaseCluster, to connect to Couchbase Server. We shall connect using the CouchbaseJavaClient application. The CouchbaseCluster class is the main entry point for connecting to the Couchbase Server. The CouchbaseCluster class provides the overloaded, static create() method to create an instance of CouchbaseCluster constructors, which are listed in Table 2-2.

***Table 2-2.*** *CouchbaseCluster Class Constructors*

| Constructor | Description |
| --- | --- |
| create() | Creates a CouchbaseCluster instance. |
| create(CouchbaseEnvironment environment) | Creates a CouchbaseCluster instance using a CouchbaseEnvironment. CouchbaseEnvironment represents the Couchbase properties such as connection timeout, disconnect timeout, query timeout, and view timeout shared across the cluster. |
| create(CouchbaseEnvironment environment, java.util.List<java.lang.String> nodes) | Creates a CouchbaseCluster instance using a CouchbaseEnvironment and a List of nodes. |
| create(CouchbaseEnvironment environment, java.lang.String... nodes) | Same as the preceding method except that the second parameter is a vararg of type String instead of a List. |
| create(java.util.List<java.lang.String> nodes) | Creates a CouchbaseCluster using a List of nodes. |
| create(java.lang.String... nodes) | Same as the preceding method except that the parameter s a String vararg instead of a List. |

Next, we connect to Couchbase cluster. First, create a CouchbaseCluster instance in the main() method using the static method create().

```
Cluster cluster = CouchbaseCluster.create();
```

Or, one of the overloaded create methods may be used to create a CouchbaseCluster instance. For example, to create a CouchbaseCluster instance using a List of nodes use the create(java.util. List<java.lang.String> nodes) method.

```
List<String> nodes = Arrays.asList("192.168.1.71", "192.168.1.72");
cluster = CouchbaseCluster.create(nodes);
```

The CouchbaseCluster provides the overloaded openBucket() method, which returns a Bucket instance, to connect to a Couchbase bucket. Invoke the openBucket(java.lang.String name, java.lang.String password) method to connect to the "json" bucket created in the previous section.

```
Bucket jsonBucket = cluster.openBucket("json", "password");
```

Alternatively, connect to the default bucket using the openBucket() method. For connecting to the default bucket, bucket name and password are not required to be specified.

```
Bucket defaultBucket = cluster.openBucket();
```

The Bucket interface represents a connection to a bucket to perform operations on the bucket synchronously. An asynchronous representation of a bucket is also provided with the AsyncBucket interface, which may be used to perform asynchronous operations on a bucket. An AsyncBucket instance may be created as follows.

```
AsyncBucket asyncBucket = jsonBucket.async();
```

To disconnect from all open buckets invoke the disconnect() method.

```
cluster.disconnect();
```

The starter CouchbaseJavaClient class is listed below.

```
import com.couchbase.client.CouchbaseClient;
import java.io.IOException;
import java.net.URI;
import java.util.LinkedList;
import java.util.List;
import java.util.concurrent.TimeUnit;
package couchbase;

import java.util.Arrays;
import java.util.List;

import com.couchbase.client.java.AsyncBucket;
import com.couchbase.client.java.Bucket;
import com.couchbase.client.java.Cluster;
import com.couchbase.client.java.CouchbaseCluster;

public class CouchbaseJavaClient {

    private static Bucket jsonBucket;

    public static void main(String args[]) {
        Cluster cluster = CouchbaseCluster.create();
        //List<String> nodes = Arrays.asList("192.168.1.71", "192.168.1.72");
        // cluster = CouchbaseCluster.create(nodes);
        jsonBucket = cluster.openBucket("json", "password");
        Bucket defaultBucket = cluster.openBucket();
        AsyncBucket asyncBucket = jsonBucket.async();
        cluster.disconnect();
    }
    public static void createDocument() {
    }
    public static void getDocument() {
    }
    public static void updateDocument() {
    }
    public static void removeDocument() {
    }
    public static void queryView() {
    }
}
```

31

Before we may run the `CouchbaseJavaClient.java` application we need to install the Maven application. Right-click on `pom.xml` and select Run As ➤ Maven; install as shown in Figure 2-13.

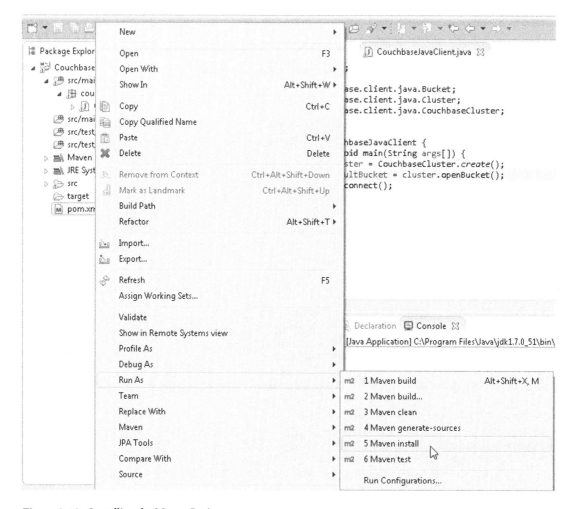

***Figure 2-13.*** *Installing the Maven Project*

Maven project gets built as shown in Figure 2-14. The Maven project is required to be built only once before the first time the Java application `CouchbaseJavaClient` is run. For subsequent modifications and runs, the Maven project is not required to be rebuilt and reinstalled.

*Figure 2-14. Maven Project Built*

Subsequently, right-click on CouchbaseJavaClient.java source file in Package Explorer and select Run As ➤ Java Application as shown in Figure 2-15.

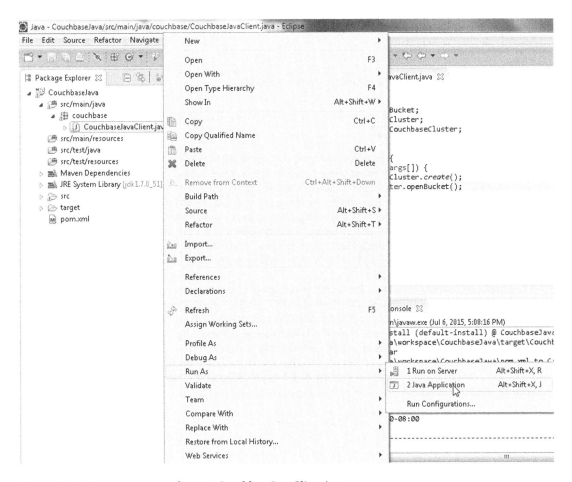

*Figure 2-15. Running Java Application CouchbaseJavaClient.java*

The client gets connected to the Couchbase Server. The output from running the class shows that the client "Connected to Node 127.0.0.1" and subsequently "Opened bucket default," "Closed bucket default," and "Disconnected from Node 127.0.0.1 as shown in Figure 2-16.

```
Problems  @ Javadoc  Declaration  Console ⊠                                          X  %     ⬚ ⬚ ⬚  ⬚ ⬚ ▾ ⬚ ▾ ⬚ ⬚
<terminated> CouchbaseJava [Java Application] C:\Program Files\Java\jdk1.7.0_51\bin\javaw.exe (Jul 6, 2015, 5:09:14 PM)
Jul 06, 2015 5:09:17 PM com.couchbase.client.core.env.DefaultCoreEnvironment <init>
INFO: ioPoolSize is less than 3 (2), setting to: 3
Jul 06, 2015 5:09:17 PM com.couchbase.client.core.env.DefaultCoreEnvironment <init>
INFO: computationPoolSize is less than 3 (2), setting to: 3
Jul 06, 2015 5:09:17 PM com.couchbase.client.core.CouchbaseCore <init>
INFO: CouchbaseEnvironment: {sslEnabled=false, sslKeystoreFile='null', sslKeystorePassword='null', queryEnabled=false, queryPort=8093, bootstrapH
, bootstrapCarrierEnabled=true, bootstrapHttpDirectPort=8091, bootstrapHttpSslPort=18091, bootstrapCarrierDirectPort=11210, bootstrapCarrierSslPc
lSize=3, computationPoolSize=3, responseBufferSize=16384, requestBufferSize=16384, kvServiceEndpoints=1, viewServiceEndpoints=1, queryServiceEndp
=NioEventLoopGroup, coreScheduler=CoreScheduler, eventBus=DefaultEventBus, packageNameAndVersion=couchbase-java-client/2.1.3 (git: 2.1.3), dcpEna
ryStrategy=BestEffort, maxRequestLifetime=75000, retryDelay=ExponentialDelay{growBy 1.0 MICROSECONDS; lower=100, upper=100000}, reconnectDelay=Ex
growBy 1.0 MILLISECONDS; lower=32, upper=4096}, observeIntervalDelay=ExponentialDelay{growBy 1.0 MICROSECONDS; lower=10, upper=100000}, keepAlive
autoreleaseAfter=2000, bufferPoolingEnabled=true, queryTimeout=75000, viewTimeout=75000, kvTimeout=2500, connectTimeout=5000, disconnectTimeout=
abled=false}
Jul 06, 2015 5:09:18 PM com.couchbase.client.core.node.CouchbaseNode$1 call
INFO: Connected to Node 127.0.0.1
Jul 06, 2015 5:09:19 PM com.couchbase.client.core.config.DefaultConfigurationProvider$6 call
INFO: Opened bucket default
Jul 06, 2015 5:09:19 PM com.couchbase.client.core.config.DefaultConfigurationProvider$9 call
INFO: Closed bucket default
Jul 06, 2015 5:09:19 PM com.couchbase.client.core.node.CouchbaseNode$1 call
INFO: Disconnected from Node 127.0.0.1
```

***Figure 2-16.*** *Running Java Application CouchbaseJavaClient.java*

# Creating a Document

In this section we shall add a document to a data bucket. A document in Couchbase Server is a key/value pair in which the value is a JSON document. The com.couchbase.client.java.document.json.JsonObject class represents a JSON object that may be stored in Couchbase Server. A document is represented with the Document interface and several class implementations are provided including the JsonDocument, which creates a document from a JsonObject. The JsonObject class provides two static methods, empty() and create(), to create an empty JsonObject instance. The JsonObject class also provides the overloaded put methods to put field/value pairs in a JsonObject instance. The field name in each of these methods is of type String. A put method is provided for each of the value types String, int, long, double, boolean, JsonObject, JsonArray, and Object.

The CouchbaseCluster class provides several overloaded insert and upsert methods to add a document to a bucket. A document is represented with the Document interface. The difference between the insert and the upsert methods is that the insert methods adds a document only if the document with the specified key does not already exist while the upsert methods add a document even if a document with the same key exists. If the document already exists, a DocumentAlreadyExistsException is generated with the insert methods and a new document is not added. The upsert methods replace the document if it already exists and add a new document if it does not exist. The insert and upsert methods return an instance of Document. Some of the overloaded insert methods are discussed in Table 2-3.

**Table 2-3.** *CouchbaseCluster Class insert Methods*

| Method | Description |
|---|---|
| insert(D document) | Inserts a document if it does not already exist with the default key/value timeout. |
| insert(D document, long timeout, java.util.concurrent.TimeUnit timeUnit) | Inserts a document if it does not already exist with a custom key/value timeout. |
| insert(D document, PersistTo persistTo) | Inserts a document if it does not already exist with the default key/value timeout as the specified number of disk copies. |
| insert(D document, PersistTo persistTo, ReplicateTo replicateTo) | Inserts a document if it does not already exist with the default key/value timeout and with the specified replication. |
| insert(D document, PersistTo persistTo, ReplicateTo replicateTo, long timeout, java.util.concurrent.TimeUnit timeUnit) | Inserts a document if it does not already exist with a custom key/value timeout and the specified durability and replication. |

In the createDocument() custom method in the CouchbaseJavaClient class create a JsonObject instance for a JSON document with fields journal, publisher and edition with String values using the put(java.lang.String name, java.lang.String value) method. First, invoke the empty() method to return a JsonObject instance and subsequently invoke the put(java.lang.String name, java.lang.String value) method to add field/value pairs.

```
JsonObject catalogObj = JsonObject.empty().put("journal", "Oracle Magazine").
put("publisher", "Oracle Publishing").put("edition", "March April 2013");
```

An instance of Bucket was created earlier.

```
Cluster cluster = CouchbaseCluster.create();
Bucket defaultBucket = cluster.openBucket("default");
```

Next, we shall use the insert(D document) method in Bucket to add a document. The JsonDocument class provides several static methods, some of which are discussed in Table 2-4, to create an instance of JsonDocument.

**Table 2-4.** *JsonDocument Class Methods*

| Method | Description |
|---|---|
| empty() | Creates an empty JsonDocument instance. |
| create(java.lang.String id) | Creates an empty JsonDocument instance with the specified document id, which is unique within a bucket. |
| create(java.lang.String id, JsonObject content) | Creates a JsonDocument instance using the specified document id and JsonObject data. |
| create(java.lang.String id, JsonObject content, long cas) | Creates a JsonDocument instance using the specified document id, JsonObject data, and CAS value. |

*(continued)*

35

***Table 2-4.*** (*continued*)

| Method | Description |
|---|---|
| create(java.lang.String id, int expiry, JsonObject content, long cas) | Creates a JsonDocument instance using the specified document id, document expiry, JsonObject data, and CAS value. |
| from(JsonDocument doc, JsonObject content) | Creates a JsonDocument instance from another JsonDocument instance by replacing the JsonObject data. |
| from(JsonDocument doc, long cas) | Creates a JsonDocument from another JsonDocument instance by replacing the CAS value. |
| from(JsonDocument doc, java.lang.String id) | Creates a JsonDocument from another JsonDocument instance by replacing the document id. |
| from(JsonDocument doc, java.lang.String id, JsonObject content) | Creates a JsonDocument from another JsonDocument instance by replacing the document id and the JsonObject data. |

Add the `JsonObject` instance created earlier to the `default` bucket using an instance of `Bucket` and the `insert(D document)` method. Create an instance of `JsonDocument` using the `JsonDocument. create(java. lang.String id, JsonObject content)` method. Specify document id as "catalog."

```
Document document = defaultBucket.insert(JsonDocument.create("catalog", catalogObj));
```

Alternatively, the overloaded `upsert()` method may be used to add a new document. The `upsert()` method adds a new document if it does not already exists and replaces a document if it already exists. Some of the overloaded `upsert` methods are discussed in Table 2-5.

***Table 2-5.*** *Bucket Class upsert() Methods*

| Method | Description |
|---|---|
| upsert(D document) | Inserts a document if it does not already exist and replaces a document if it already exists with the default key/value timeout. |
| upsert(D document, long timeout, java.util.concurrent.TimeUnit timeUnit) | Inserts a document if it does not already exist and replaces a document if it already exists with a custom key/value timeout. |
| upsert(D document, PersistTo persistTo) | Inserts a document if it does not already exist and replaces a document if it already exists with the default key/value timeout to the specified number of disk copies. |
| upsert(D document, PersistTo persistTo, ReplicateTo replicateTo) | Inserts a document if it does not already exist and replaces a document if it already exists with the default key/value timeout and with the specified replication. |
| upsert(D document, PersistTo persistTo, ReplicateTo replicateTo, long timeout, java.util.concurrent.TimeUnit timeUnit) | Inserts a document if it does not already exist and replaces a document if it already exists with a custom key/value timeout and the specified durability and replication. |

If the upsert(D document) method is used instead of the insert(D document) method, a new document may be added as follows.

```
Document document = defaultBucket.upsert(JsonDocument.create("catalog", catalogObj));
```

The createDocument() method in the CouchbaseJavaClient class is as follows with jsonBucket being a class variable of type Bucket. Invoke the createDocument() method from the main() method.

```
public static void createDocument() {
JsonObject catalogObj = JsonObject.empty().put("journal", "Oracle Magazine").
put("publisher", "Oracle Publishing")
.put("edition", "March April 2013");
JsonDocument document = defaultBucket.insert(JsonDocument.create("catalog", catalogObj));
// document = jsonBucket.upsert(JsonDocument.create("catalog",
// catalogObj));
}
```

To add a document, right-click on the CouchbaseJavaClient class with the createDocument() method added and select Run As ➤ Java Application. The document object gets added to the Couchbase document store. In the Couchbase Console select Data Buckets and click on the Documents button for the "default" bucket. The catalog document that we added is listed including its content. Click on Edit Document to display the full JSON for the catalog document as shown in Figure 2-17.

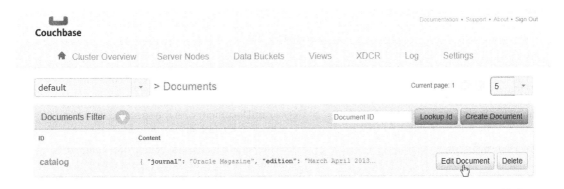

***Figure 2-17.*** *Editing/Displaying a Document in the Default Data Bucket*

The JSON for the "catalog" document gets displayed as shown in Figure 2-18.

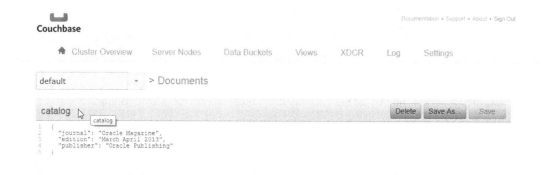

**Figure 2-18.** *Couchbase JSON Document Catalog*

# Getting a Document

In this section we shall get (or retrieve) a document from Couchbase Server in the getDocument() method of the CouchbaseJavaClient application. The Bucket class provides the overloaded get() method to get a document. Some of the get() methods are discussed in Table 2-6. Each of the get() methods returns a Document instance.

**Table 2-6.** *Bucket Class get() Methods*

| Method | Description |
| --- | --- |
| get(D document) | Gets a document with the default key/value timeout. |
| get(D document, long timeout, java.util. concurrent.TimeUnit timeUnit) | Gets a document with a custom key/value timeout. |
| get(java.lang.String id) | Gets a JsonDocument with the specified document id using the default timeout. |
| get(java.lang.String id, java.lang.Class<D> target) | Gets a document with the specified document id using the default timeout. The target type of the document is also supplied. |
| get(java.lang.String id, java.lang.Class<D> target, long timeout, java.util.concurrent.TimeUnit timeUnit) | Gets a document with the specified document id using a custom timeout. The target type of the document is also supplied. |
| get(java.lang.String id, long timeout, java.util. concurrent.TimeUnit timeUnit) | Gets a JsonDocument with the specified document id using custom timeout. |

Using an instance of Bucket for the default bucket get a JsonDocument instance with id catalog using the get(String) method.

```
JsonDocument catalog = defaultBucket.get("catalog");
```

The JsonDocument class provides various methods to get document properties as discussed in Table 2-7.

***Table 2-7.*** *JsonDocument Class Methods*

| Method | Description |
| --- | --- |
| cas() | Returns the CAS value of the document as a long. The default CAS value is 0. |
| content() | Returns document content. |
| expiry() | Returns the expiration time of the document as an int. The default value is 0. |
| id() | Returns the document id as a String. The id is unique within a bucket. |
| toString() | Returns the String representation of the document. |

Output the CAS value of the document retrieved using the cas() method and also output the document content using the content() method. The getDocument() method in CouchbaseJavaClient is as follows; catalog is a class variable of type JsonDocument.

```
public static void getDocument() {
    catalog = defaultBucket.get("catalog");
    System.out.println("Cas Value: " + catalog.cas());
    System.out.println("Catalog: " + catalog.content());
}
```

Right-click on CouchbaseJavaClient.java in Package Explorer and select Run As ➤ Java Application. The document with the specified id gets retrieved and its CAS value and content get output as shown in Figure 2-19.

```
ⓜ CouchbaseJava/pom.xml        🗊 CouchbaseJavaClient.java ⊠
 23
 24            cluster.disconnect();
 25        }
 26
 27⊖    public static void createDocument() {
 28
 29            catalogObj = JsonObject.empty()
 30                .put("journal", "Oracle Magazine")
 31                .put("publisher", "Oracle Publishing")
 32                .put("edition", "March April 2013");
 33            document = defaultBucket.upsert(JsonDocument.create("catalog", catalogObj));
 34
 35        }
 36
 37⊖    public static void getDocument() {
 38
 39            catalog = defaultBucket.get("catalog");
 40        System.out.println("Cas Value: " + catalog.cas());
 41        System.out.println("Catalog: " + catalog.content());
 42
 43        }
```

```
🔧 Problems   @ Javadoc   🔍 Declaration   🖥 Console ⊠
<terminated> CouchbaseJavaClient [Java Application] C:\Program Files\Java\jdk1.7.0_51\bin\javaw.exe (Jul 6, 2015, 5:48:44 PM)
abled=false}
Jul 06, 2015 5:48:47 PM com.couchbase.client.core.node.CouchbaseNode$1 call
INFO: Connected to Node 127.0.0.1
Jul 06, 2015 5:48:48 PM com.couchbase.client.core.config.DefaultConfigurationProvider$6 call
INFO: Opened bucket json
Cas Value: 1436229230842159204
Catalog: {"journal":"Oracle Magazine","edition":"March April 2013","publisher":"Oracle Publishing"}
Jul 06, 2015 5:48:48 PM com.couchbase.client.core.config.DefaultConfigurationProvider$9 call
INFO: Closed bucket json
Jul 06, 2015 5:48:48 PM com.couchbase.client.core.node.CouchbaseNode$1 call
INFO: Disconnected from Node 127.0.0.1
```

***Figure 2-19.*** *Getting a Document*

# Updating a Document

In this section we shall update a document previously added to a Couchbase bucket. We shall use the updateDocument() custom method in the CouchbaseJavaClient class for the update. Assuming a document has previously been added to the json bucket we need to create an instance of Bucket for the json bucket using the openBucket() method of a CouchbaseCluster instance.

```
Bucket jsonBucket = cluster.openBucket("json", "password");
```

The Bucket class provides two overloaded methods to update a document: upsert and replace. The upsert method as discussed earlier replaces a document if it exists and adds a document if it does not already exist. The replace method also replaces a document. The difference between replace and upsert is that the replace method should be used only if the document to be replaced exists. If the document to be replaced does not exist a DocumentDoesNotExistException error is generated. Some of the replace() methods are discussed in Table 2-8.

**Table 2-8.** *Overloaded replace() Methods*

| Method | Description |
| --- | --- |
| replace(D document) | Replaces a document using the default key/value timeout. |
| replace(D document, long timeout, java.util.concurrent.TimeUnit timeUnit) | Replaces a document using a custom key/value timeout. |
| replace(D document, PersistTo persistTo) | Replaces a document using the specified persistence and default key/value timeout. |
| replace(D document, PersistTo persistTo, ReplicateTo replicateTo) | Replaces a document using the specified persistence and replication, and default key/value timeout. |
| replace(D document, PersistTo persistTo, ReplicateTo replicateTo, long timeout, java.util.concurrent.TimeUnit timeUnit) | Replaces a document using the specified persistence and replication, and custom key/value timeout. |

We shall replace the following JSON document.

```
{
"journal":"Oracle Magazine",
"publisher":"Oracle Publishing",
"edition":"March April 2015"
}
```

The JSON document with id catalog in the json bucket is shown in Figure 2-20.

**Figure 2-20.** *Document to be replaced*

In the updateDocument() method create a JsonObject instance for the replacement document. As discussed earlier first create and empty JsonObject instance using the class method empty() in JsonObject and subsequently invoke the put(String,String) method to add field/value pairs to the JsonObject instance. Add a new field called section and modify the edition field.

41

```
JsonObject catalogObj = JsonObject.empty().put("journal", "Oracle Magazine").put("publisher",
"Oracle Publishing").put("edition", "January February 2015").put("section", "Technology");
```

Invoke the replace(D document) method to replace the catalog document with a replacement document.

```
jsonBucket.replace(JsonDocument.create("catalog",catalogObj));
```

Alternatively, invoke the upsert(D document) method.

```
jsonBucket.upsert(JsonDocument.create("catalog", catalogObj));
```

The updateDocument() custom method is as follows.

```
public static void updateDocument() {
    catalogObj = JsonObject.empty().put("journal", "Oracle Magazine")
    .put("publisher", "Oracle Publishing")
    .put("edition", "January February 2015")
    .put("section", "Technology");
     jsonBucket.replace(JsonDocument.create("catalog",catalogObj));
    // jsonBucket.upsert(JsonDocument.create("catalog", catalogObj));
}
```

Invoke the replaceDocument() method in the main() method and subsequently invoke the getDocument() method to get and output the new document.

```
replaceDocument();
getDocument();
```

Right-click on the CouchbaseJavaClient application and select Run As ➤ Java Application. The document with id catalog in the json bucket gets replaced, and the replaced document CAS value and JSON get output to the Eclipse console.

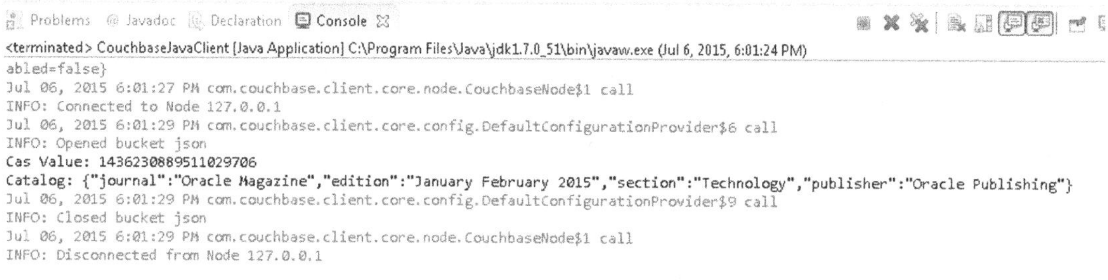

**Figure 2-21.** *Replaced Document CAS and JSON data*

The document in the Couchbase json bucket with id catalog gets replaced as shown in Figure 2-22. The edition field has been modified and a new field called section has been added.

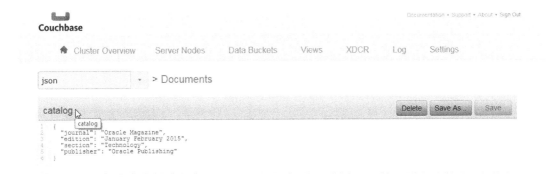

**Figure 2-22.** *Replaced Document*

Next, we shall query a document using a Couchbase View. But, first we need to create the View.

# Creating a View

The JSON data stored in Couchbase Server can be indexed using a View, which creates an index on the data according to the defined format and structure. A View extracts the fields from the JSON document object in Couchbase Server and creates an index that can be queried. A view is a logical structure, and a map function maps the fields of the JSON document object stored in the Couchbase Server to a view.

Optionally a reduce function can also be applied to summarize (or average or sum) the data. In this section we create a View on the JSON document in the Couchbase Server. A map function has the following format.

```
function(doc, meta)
{
  emit(doc.name, [doc.field1, doc.field2]);
}
```

When the function is translated to a map() function, the map() function is supplied with two arguments for each document stored in a bucket: the doc arg and the meta arg. The doc arg is the document object stored in the Couchbase bucket and its content type can be identified with the meta.type field. The meta arg is the metadata for the document object stored in the bucket. Every document in the data bucket is submitted to the map() function. Within the map() function any custom code can be specified. The emit() function is used to emit a row or a record of data from the map() function. The emit() function takes two arguments: a key and a value.

```
emit(key,value)
```

The emitted key is used for sorting and querying the document object fields mapped to the view. The key may have any format such as a string, a number, a compound structure such as an array, or a JSON object. The value is the data to be output in a row or record and it may have any format including a string, number, an array, or JSON. Specify the following function for the mapping from the Couchbase Server bucket to the view. The function first tests if the type of the document is JSON and subsequently emits records with each record key being the document name and each record value being the data stored in the fields of the document object.

Next, create a View in Couchbase Console. Select Data Buckets ➤ json bucket. Subsequently, select View. The Development View tab is selected by default. Click on Create Development View to create a development view as shown in Figure 2-23.

**Figure 2-23.** *Selecting Create Development View*

In Create Development View dialog specify a Design Document Name (_design/dev_catalog) and View Name (catalog_view) as shown in Figure 2-24. The _design prefix is not included in the design document name when accessed programmatically with a Java client. Click on Save.

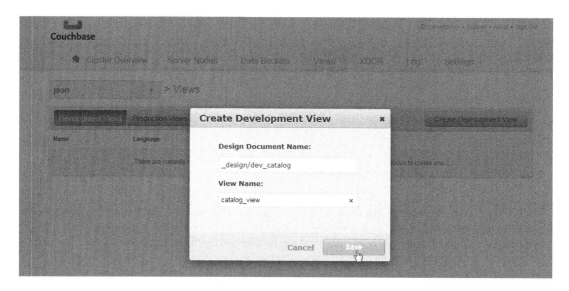

**Figure 2-24.** *Creating a Development View*

A development view called catalog_view gets created as shown in Figure 2-25.

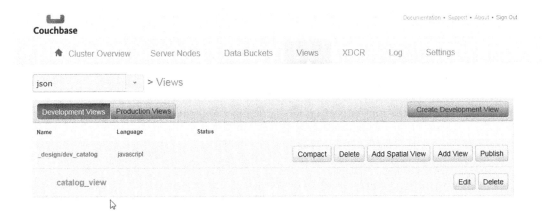

***Figure 2-25.*** *Development View catalog_view*

We need to convert the development view to a production view before we are able to access the view from a Couchbase Java client. Click on Publish as shown in Figure 2-26 to convert the view to a production view.

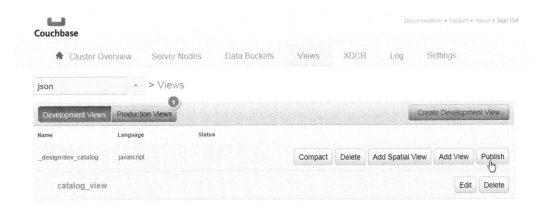

***Figure 2-26.*** *Converting a Development View to Production View*

The Production View gets created as shown in Figure 2-27.

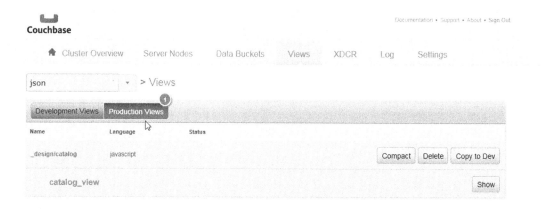

**Figure 2-27.** *Production View*

Next, add a document to be indexed by the view to the json bucket. Select the json bucket and click on Documents. Click on Create Document as shown in Figure 2-28.

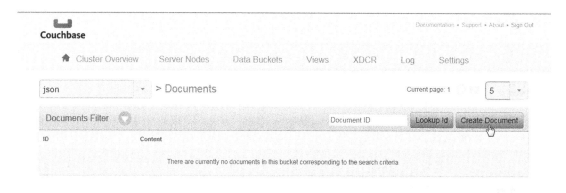

**Figure 2-28.** *Creating a Document in json Bucket*

In Create Document dialog specify Document Id as catalog and click on Create as shown in Figure 2-29.

**Figure 2-29.** *Specifying a Document Id*

A new document with catalog id gets created. Copy and paste the following JSON document to the catalog id document JSON and click on Save.

```
{
"journal":"Oracle Magazine",
"publisher":"Oracle Publishing",
"edition":"March April 2015"
}
```

A JSON document gets added to the json bucket as shown in Figure 2-30.

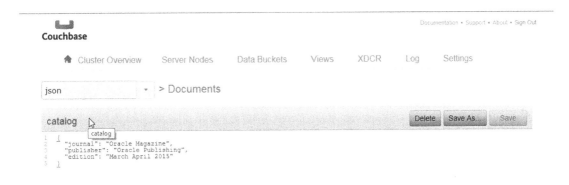

**Figure 2-30.** *Adding JSON document*

47

We had previously added the catalog_view to the json bucket. To index the catalog id document using the catalog_view view select the catalog_view view in json ➤ Views. Copy the following map function to View Code ➤ Map.

```
function(doc,meta) {    if (meta.type == 'json') {    emit(doc.name,
[doc.journal,doc.publisher,doc.edition]);    } }
```

The catalog_view is shown in Figure 2-31 with the catalog document indexed using the specified Map function.

***Figure 2-31.***  *Mapping the catalog document using a View Map Function*

# Querying a View

Next, we shall query the view we created in the previous section. A View query is represented with the ViewQuery class, which provides the from(java.lang.String design, java.lang.String view) class method to create a ViewQuery instance. The Bucket class provides the overloaded query() method discussed in Table 2-9 to query a view. Each of the query() methods return a ViewResult instance, which represents the result from a ViewQuery.

***Table 2-9.*** *Overloaded query() Methods*

| Method | Description |
|--------|-------------|
| query(ViewQuery query) | Queries a Couchbase Server View with the default timeout. |
| query(ViewQuery query, long timeout, java.util.concurrent.TimeUnit timeUnit) | Queries a Couchbase Server View with a custom timeout. |

In the `queryView()` method invoke the `query(ViewQuery query)` method in Bucket. Create a ViewQuery argument using the `static` method `from(java.lang.String design, java.lang.String view)` with the design document name as "catalog" and view name as "catalog_view," which were created in the preceding section.

```
ViewResult result = jsonBucket.query(ViewQuery.from("catalog","catalog_view"));
```

`ViewResult` provides the overloaded `rows()` method that returns an `Iterator` over the rows in the view result. The `ViewRow` interface represents a view row. Using an enhanced `for` loop, iterate over the rows in the `ViewResult` and output the row value for each row.

```
for (ViewRow row : result) {
System.out.println(row);
}
```

The `queryView()` custom method is as follows.

```
public static void queryView() {
    ViewResult result = jsonBucket.query(ViewQuery.from("catalog","catalog_view"));
    for (ViewRow row : result) {
        System.out.println(row);
    }
}
```

Invoke the `queryView()` method in the `main()` method. When the `CouchbaseJavaClient` class is run, the output from the view query shown in Figure 2-32 gets displayed.

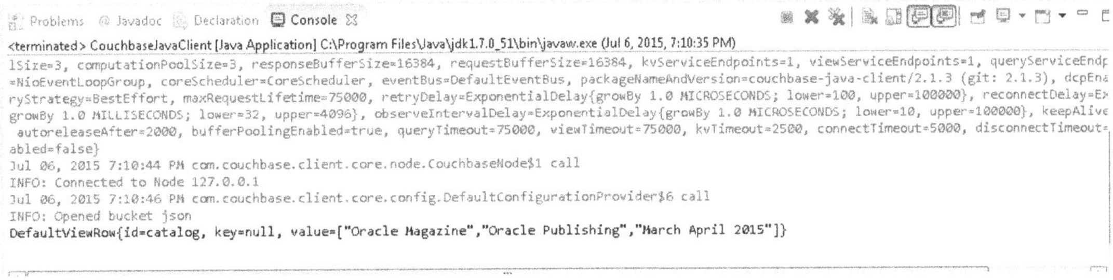

***Figure 2-32.*** *Result of Querying a View*

# Deleting a Document

In this section we shall delete a document from Couchbase Server.

The Bucket class provides the overloaded remove() method to remove a document. Some of the remove() methods are discussed in Table 2-10.

**Table 2-10.** *Overloaded remove() Methods*

| Method | Description |
| --- | --- |
| remove(D document) | Removes a document using the default key/value timeout. Returns a Document instance. |
| remove(D document, long timeout, java.util.concurrent.TimeUnit timeUnit) | Removes a document using a custom key/value timeout. Returns a Document instance. |
| remove(D document, PersistTo persistTo, ReplicateTo replicateTo, long timeout, java.util.concurrent.TimeUnit timeUnit) | Removes a document using the specified persistence and replication and default key/value timeout. Returns a Document instance. |
| remove(java.lang.String id) | Removes a document using the specified document id. Returns a JsonDocument instance. |
| remove(java.lang.String id, PersistTo persistTo, ReplicateTo replicateTo, long timeout, java.util.concurrent.TimeUnit timeUnit) | Removes a document using the specified document id, persistence and replication, and custom key/value timeout. Returns a JsonDocument instance. |

In the removeDocument() custom method invoke the remove(String id) method in a Bucket instance to remove the catalog id document from the json bucket. Subsequently output the CAS value and document JSON data.

```java
public static void removeDocument() {
    document = jsonBucket.remove("catalog");
    System.out.println("Cas Value: " + document.cas());
    System.out.println("Catalog: " + document.content());
}
```

The CoucbaseJavaClient class used to add, get, update, query using a view, and delete a document in Couchbase Server, and query a view is listed.

```java
package couchbase;

import java.util.ArrayList;
import java.util.Arrays;
import java.util.List;

import com.couchbase.client.java.AsyncBucket;
import com.couchbase.client.java.Bucket;
import com.couchbase.client.java.Cluster;
import com.couchbase.client.java.CouchbaseCluster;
import com.couchbase.client.java.document.JsonDocument;
import com.couchbase.client.java.document.json.JsonObject;
```

```java
import com.couchbase.client.java.view.DefaultView;
import com.couchbase.client.java.view.DesignDocument;
import com.couchbase.client.java.view.View;
import com.couchbase.client.java.view.ViewQuery;
import com.couchbase.client.java.view.ViewResult;
import com.couchbase.client.java.view.ViewRow;

public class CouchbaseJavaClient {

    private static Bucket jsonBucket;
    private static JsonDocument document;
    private static JsonObject catalogObj;
    private static JsonDocument catalog;

    public static void main(String args[]) {
        Cluster cluster = CouchbaseCluster.create();
        //jsonBucket = cluster.openBucket("json", "calgary10");

        // AsyncBucket asyncBucket = jsonBucket.async();

        // List<String> nodes = Arrays.asList("192.168.1.71", "192.168.1.72");
        // cluster = CouchbaseCluster.create(nodes);

        // Bucket defaultBucket = cluster.openBucket();

        // createDocument();
        // updateDocument();
        // getDocument();
        // removeDocument();
        //queryView();

        //cluster.disconnect();
    }

    public static void createDocument() {

        catalogObj = JsonObject.empty().put("journal", "Oracle Magazine")
                .put("publisher", "Oracle Publishing")
                .put("edition", "March April 2013");

        document = defaultBucket
                .insert(JsonDocument.create("catalog", catalogObj));
        // document = defaultBucket.upsert(JsonDocument.create("catalog",
        // catalogObj));

    }
```

```java
    public static void getDocument() {

        catalog = defaultBucket.get("catalog");
        System.out.println("Cas Value: " + catalog.cas());
        System.out.println("Catalog: " + catalog.content());

    }

    public static void updateDocument() {

        catalogObj = JsonObject.empty().put("journal", "Oracle Magazine")
                .put("publisher", "Oracle Publishing")
                .put("edition", "January February 2015")
                .put("section", "Technology");
         jsonBucket.replace(JsonDocument.create("catalog",catalogObj));
        // jsonBucket.upsert(JsonDocument.create("catalog", catalogObj));

    }

    public static void removeDocument() {

        document = jsonBucket.remove("catalog");
        System.out.println("Cas Value: " + document.cas());
        System.out.println("Catalog: " + document.content());

    }

    public static void queryView() {

        ViewResult result = jsonBucket.query(ViewQuery.from("catalog","catalog_view"));
        for (ViewRow row : result) {
            System.out.println(row);
        }

    }
}
```

Right-click on the CouchbaseJavaClient Java source file in Package Explorer and select Run As ➤ Java Application. The output from the application is shown in Figure 2-33.

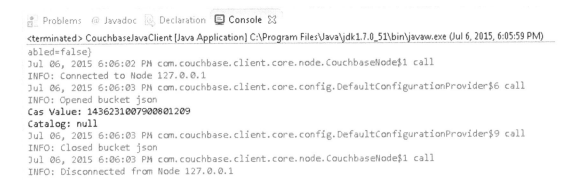

*Figure 2-33. Result of Running the CouchbaseJavaClient Application to Delete a Document*

# Summary

In this chapter we learned to add, get, update, and delete a document in a Couchbase Server bucket with the Couchbase Java Client library. We also created a view for a document and queried the document using a view query with the Couchbase Java Client library. In the next chapter we shall discuss using Couchbase Server with Spring Data.

# CHAPTER 3

■ ■ ■

# Using Spring Data

Spring Data is designed for new data access technologies such as non-relational databases. Couchbase is a non-relational NoSQL database with benefits such as scalability, flexibility, and high performance. The Spring Data Couchbase project adds Spring Data functionality to the Couchbase Server. This chapter explains how to use the Spring Data Couchbase project in Eclipse and assumes knowledge about using Maven. The chapter covers the following topics.

- Setting Up the Environment
- Creating a Maven Project
- Installing Spring Data Couchbase
- Configuring JavaConfig
- Creating a Model
- Using Spring Data with Couchbase with Template
- Running Couchbase CRUD Operations
- Using Spring Data Repositories with Couchbase

## Setting Up the Environment

Download and install the following software components.

- Couchbase Server 3.0.x for Windows, Enterprise Edition from http://www.couchbase.com/nosql-databases/downloads.
- Eclipse IDE for Java EE Developers from http://www.eclipse.org/downloads/.
- JDK 1.5 or later from http://www.oracle.com/technetwork/java/javase/downloads/jdk7-downloads-1880260.html.

## Creating a Maven Project

First, we need to create a Maven project in Eclipse.

1. Select File ➤ New ➤ Other. In the New window, select the Maven ➤ Maven Project wizard as shown in Figure 3-1 and click on Next.

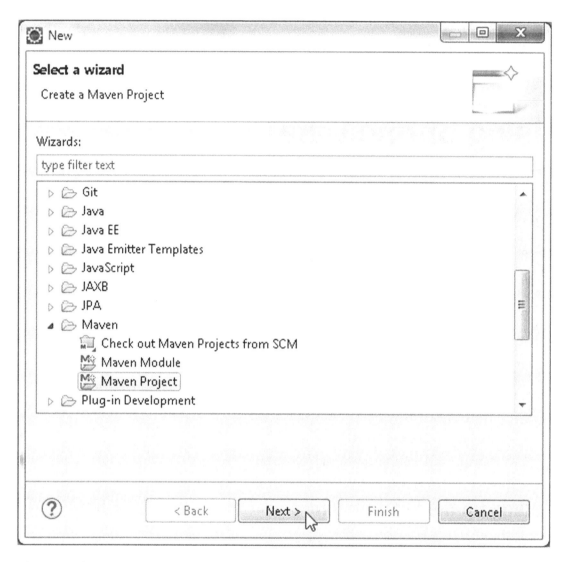

**Figure 3-1.** *Creating a new Maven Project*

2.  The New Maven Project wizard gets started. Select the Create a simple project
    check box and the Use default Workspace location check box as shown in
    Figure 3-2. Click on Next.

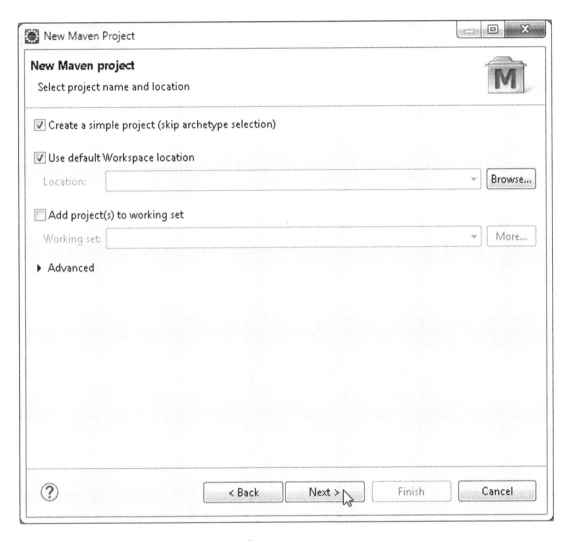

***Figure 3-2.*** *Selecting Maven Project Name and Location*

3.  In Configure project specify the following as shown in Figure 3-3 and click on Finish.

    - Group Id (com.couchbase.core)

    - Artifact Id (SpringCouchbase)

    - Version (1.0.0)

    - Packaging (jar)

    - Name (SpringCouchbase)

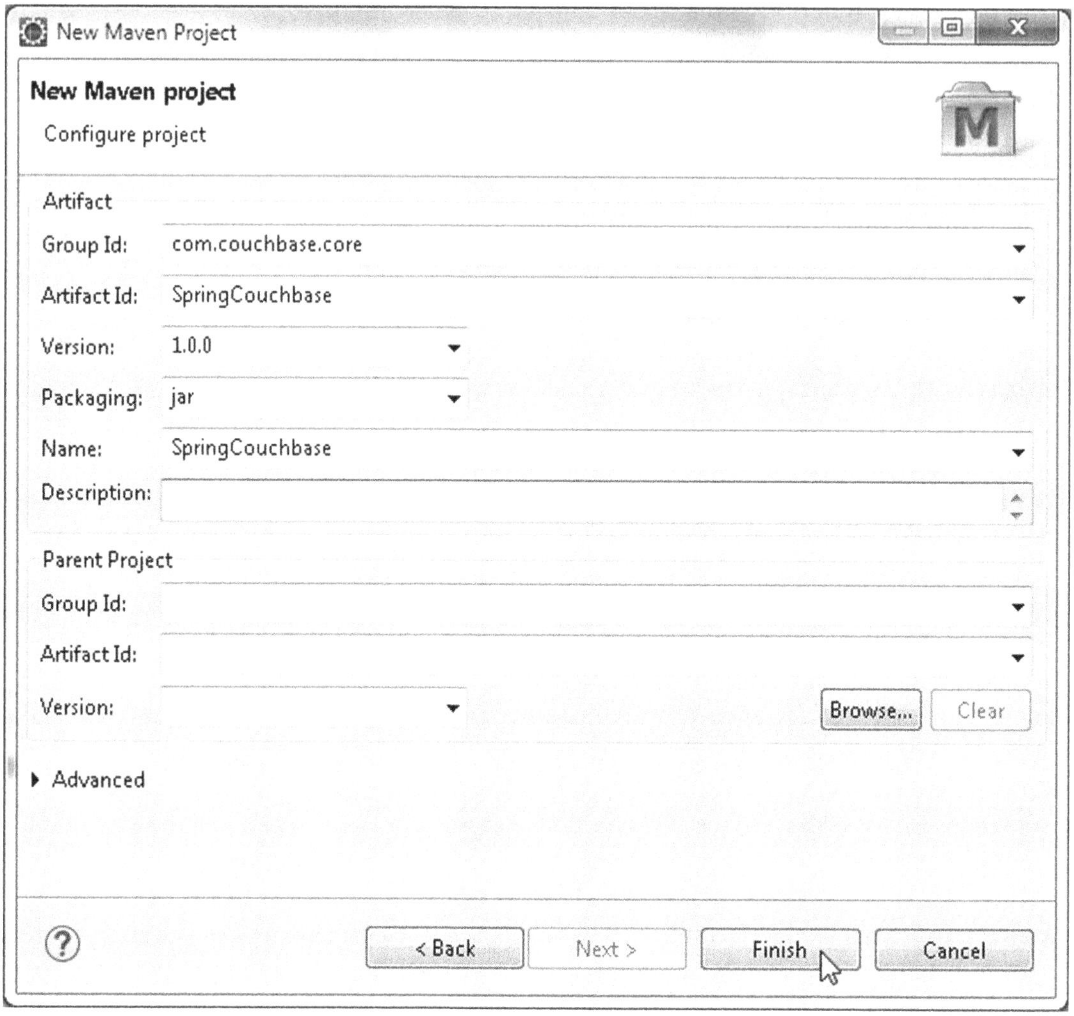

*Figure 3-3.* *Configuring Maven Project*

A Maven project (SpringCouchbase) gets created as shown in Package Explorer in Figure 3-4. A newly created project could have error markers, which would get fixed as the application is developed.

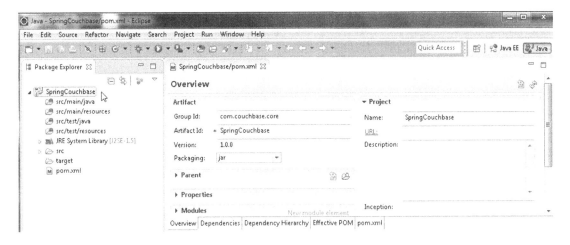

**Figure 3-4.** *Maven Project SpringCouchbase*

We also need to add some Java classes to the Maven project to demonstrate the use of Spring Data Couchbase. Add the Java classes listed in Table 3-1.

**Table 3-1.** *Java Classes*

| Class | Description |
| --- | --- |
| com.couchbase.config.<br>SpringCouchbaseApplicationConfig | JavaConfig class. |
| com.couchbase.core.App | Java application for using Spring Data with Couchbase with Template. |
| com.couchbase.model.Catalog | Model class. |
| com.couchbase.repositories.CatalogRepository | Implementation class for Couchbase-specific Repository. |
| com.couchbase.service.CatalogService | Service class to invoke CRUD operations on Couchbase Repository. |

The Java classes are shown in Figure 3-5.

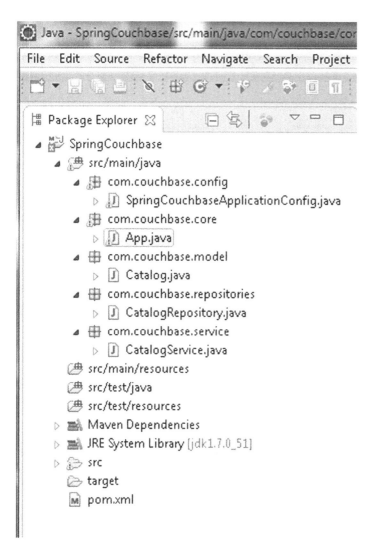

*Figure 3-5.* *Java Classes in Maven Project*

# Installing Spring Data Couchbase

The Maven project includes a pom.xml to specify the dependencies for the project and the build configuration for the project. Specify the dependency/(ies) listed in Table 3-2 in pom.xml.

*Table 3-2.* *Maven Project Dependencies*

| Dependency Group Id | Artifact Id | Version | Description |
|---|---|---|---|
| org.springframework.data | spring-data-couchbase | 1.3.1.RELEASE | Spring Data Couchbase |

Specify the `maven-compiler-plugin` and `maven-eclipse-plugin` plug-in in the build configuration. The `pom.xml` to use the Spring Data Couchbase project is listed.

```
<project xmlns="http://maven.apache.org/POM/4.0.0" xmlns:xsi="http://www.w3.org/2001/
XMLSchema-instance"
    xsi:schemaLocation="http://maven.apache.org/POM/4.0.0 http://maven.apache.org/xsd/
    maven-4.0.0.xsd">
    <modelVersion>4.0.0</modelVersion>
    <groupId>com.couchbase.core</groupId>
    <artifactId>SpringCouchbase</artifactId>
    <version>1.0.0</version>
    <name>SpringCouchbase</name>
    <dependencies>
        <dependency>
            <groupId>org.springframework.data</groupId>
            <artifactId>spring-data-couchbase</artifactId>
            <version>1.3.1.RELEASE</version>
        </dependency>
    </dependencies>
    <build>
        <plugins>
            <plugin>
                <artifactId>maven-compiler-plugin</artifactId>
                <version>3.0</version>
                <configuration>
                    <source>1.7</source>
                    <target>1.7</target>
                </configuration>
            </plugin>
            <plugin>
                <groupId>org.apache.maven.plugins</groupId>
                <artifactId>maven-eclipse-plugin</artifactId>
                <version>2.9</version>
                <configuration>
                    <downloadSources>true</downloadSources>
                    <downloadJavadocs>true</downloadJavadocs>
                </configuration>
            </plugin>
        </plugins>
    </build>
</project>
```

The Spring Data Couchbase dependency has sub-dependencies, which also get downloaded when the pom.xml is saved with File ➤ Save All. The jars for all the dependencies get added to the Java Build Path of the Maven project as shown in Figure 3-6.

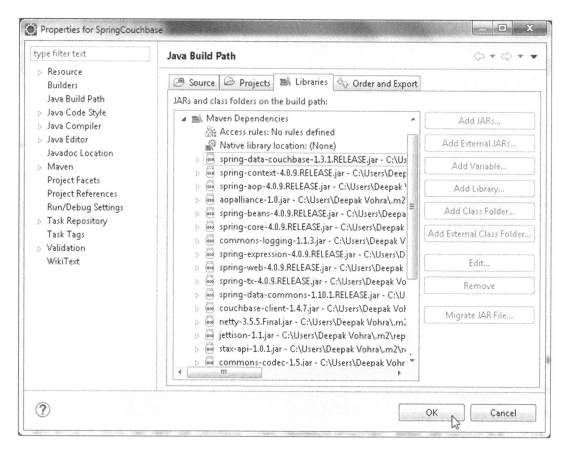

***Figure 3-6.*** *Maven Project SpringCouchbase Java Build Path*

# Configuring JavaConfig

In this section we shall configure the Spring environment with POJOs using JavaConfig, which is known officially as "annotation-based configuration." The base class for Spring Data Couchbase configuration with JavaConfig is `org.springframework.data.couchbase.config.AbstractCouchbaseConfiguration`.

1. Create a class, `SpringCouchbaseApplicationConfig`, which declares some `@Bean` methods and extends the `org.springframework.data.couchbase` `.config.AbstractCouchbaseConfiguration` class.

2. Annotate the class with `@Configuration`, which indicates that the class is processed by the Spring container to generate bean definitions and service requests for the beans at runtime.

3. Declare a `@Bean` annotated method that returns a `CouchbaseClient` instance.

4. The `SpringCouchbaseApplicationConfig` class must implement the inherited abstract method `couchbaseClient()`.

5. The `http://127.0.0.1:8091/pools` URI is used to create a `CouchbaseClient` instance for the "default" bucket, which does not require a password.

The Spring configuration class SpringCouchbaseApplicationConfig is listed below.

```java
package com.couchbase.config;

import org.springframework.context.annotation.Bean;
import org.springframework.context.annotation.Configuration;
import org.springframework.data.couchbase.config.AbstractCouchbaseConfiguration;
import org.springframework.data.couchbase.repository.config.EnableCouchbaseRepositories;

import com.couchbase.client.CouchbaseClient;

import java.net.URI;
import java.util.Arrays;

import java.util.List;

@Configuration
//@EnableCouchbaseRepositories("com.couchbase.repositories")
public class SpringCouchbaseApplicationConfig extends
    AbstractCouchbaseConfiguration {
    @Bean
    public CouchbaseClient couchbaseClient() throws Exception {
        return new CouchbaseClient(Arrays.asList(new URI(
                "http://127.0.0.1:8091/pools")), "default", "");
    }

    @Override
    protected List<String> bootstrapHosts() {

        return Arrays.asList(new String("http://127.0.0.1:8091/pools"));
    }

    @Override
    protected String getBucketName() {

        return "default";
    }

    @Override
    protected String getBucketPassword() {

        return "";
    }
}
```

# Creating a Model

Next, we'll create the model class to use with the Spring Data Couchbase project. A domain object to be persisted to Couchbase Server must be annotated with @Document.

1. Create a POJO class Catalog with fields for id, journal, edition, publisher, title, and author and the corresponding get/set methods.

2. Annotate the id field with @Id.

3. Add a constructor that may be used to construct a Catalog instance.

The Catalog entity is listed below.

```
package com.couchbase.model;
import org.springframework.data.annotation.Id;
import org.springframework.data.couchbase.core.mapping.Document;

@Document
public class Catalog {

  @Id
  private String id;
  private String journal;
  private String publisher;
  private String edition;
  private String title;
  private String author;
public String getId() {
    return id;
  }
public void setId(String id) {
    this.id = id;
  }
public String getJournal() {
    return journal;
  }
public void setJournal(String journal) {
    this.journal = journal;
  }
public String getPublisher() {
    return publisher;
  }
public void setPublisher(String publisher) {
    this.publisher = publisher;
  }
public String getEdition() {
    return edition;
  }
public void setEdition(String edition) {
    this.edition = edition;
  }
```

```
public String getTitle() {
     return title;
   }
public void setTitle(String title) {
     this.title = title;
   }
public String getAuthor() {
     return author;
   }
public void setAuthor(String author) {
     this.author = author;
   }
  public Catalog(String journal, String publisher, String edition, String title, String
  author) {
    id = "catalog:" + title.toLowerCase().replace(" ", "-");
    this.journal = journal;
    this.publisher = publisher;
    this.edition = edition;
    this.title = title;
    this.author = author;
  }
```

Add class methods listed in Table 3-3 to the App.java application. In subsequent sections we shall run various CRUD operations using these class methods.

***Table 3-3.*** *Class Methods*

| Method | Description |
| --- | --- |
| saveDocument() | Saves a single document |
| saveDocuments() | Saves multiple documents |
| removeDocument() | Removes a single document |
| removeDocuments() | Removes multiple documents |
| insertDocument() | Inserts a single document |
| insertDocuments() | Inserts a collection of documents |
| documentExists() | Finds if a document exists |
| findDocumentById() | Finds document by Id |
| findDocumentByView() | Finds document by view |
| queryDocumentView() | Query Document View |
| updateDocument() | Update document |
| updateDocuments() | Update documents |
| bucketCallback() | Bucket Callback |

# Using Spring Data with Couchbase with Template

The common CRUD operations on a Couchbase datasource may be performed using the org.springframework.data.couchbase.core.CouchbaseOperations interface. The org.springframework .data.couchbase.core.CouchbaseTemplate class implements the CouchbaseOperations interface. A CouchbaseTemplate instance may be obtained using the ApplicationContext. In the com.couchbase.core. App application, which we created earlier, create an ApplicationContext as follows.

```
ApplicationContext context = new AnnotationConfigApplicationContext
(SpringCouchbaseApplicationConfig.class);
```

The getBean(String name,Class requiredType) method returns a named bean of the specified type. The bean name for a CouchbaseTemplate is couchbaseTemplate. The class type is CouchbaseOperations.class.

```
CouchbaseOperations ops = context.getBean("couchbaseTemplate",CouchbaseOperations.class);
```

The CouchbaseOperations instance may be used to perform various CRUD operations on a domain object stored in the Couchbase. For example, create a Catalog instance and save it in the Couchbase Server using the save(Object objectToSave) method.

```
Catalog catalog1 = new Catalog("Oracle Magazine", "Oracle Publishing",
"November-December 2013", "Engineering as a Service","David A. Kelly");
ops.save(catalog1);
```

Create another Catalog instance and persist it to Couchbase Server.

```
Catalog catalog2 = new Catalog("Oracle Magazine", "Oracle Publishing",
"November-December 2013", "Quintessential and Collaborative","Tom Haunert");
ops.save(catalog2);
```

Subsequently, the Id of the domain object stored may be output using the getId() method.

```
System.out.println("Catalog ID : " + catalog1.getId());
System.out.println("Catalog ID : " + catalog2.getId());
```

The application used to create a CouchbaseOperations/CouchbaseTemplate instance is listed below. Some method definitions do not have any code in them as we shall further develop the application in subsequent sections. Method invocations for each of the class methods have been added to the main method. Uncomment the method invocation that is to be invoked before running the App application. For example, to run the App application for the saveDocument() method uncomment the saveDocument() method invocation in the main method and run the App application.

```
package com.couchbase.core;

import org.springframework.context.ApplicationContext;
import org.springframework.context.annotation.AnnotationConfigApplicationContext;
import com.couchbase.config.SpringCouchbaseApplicationConfig;
import org.springframework.data.couchbase.core.CouchbaseOperations;
import com.couchbase.model.Catalog;
```

```java
public class App {
    static CouchbaseOperations ops;
    static Catalog catalog1;
    static Catalog catalog2;

    public static void main(String[] args) {
        ApplicationContext context = new AnnotationConfigApplicationContext(
                SpringCouchbaseApplicationConfig.class);
        ops = context.getBean("couchbaseTemplate", CouchbaseOperations.class);
        catalog1 = new Catalog("Oracle Magazine", "Oracle Publishing",
                "November-December 2013", "Engineering as a Service",
                "David A. Kelly");
        catalog2 = new Catalog("Oracle Magazine", "Oracle Publishing",
                "November-December 2013", "Quintessential and Collaborative",
                "Tom Haunert");

          saveDocument();
        // saveDocuments();
        // removeDocument();
        // removeDocuments();
        // insertDocument();
        // insertDocuments();
        // documentExists();
        // findDocumentById();
        // findDocumentByView();
        // queryDocumentView();
        // updateDocument();
        // updateDocuments();
        //bucketCallback();
    }

    public static void saveDocument() {
        ops.save(catalog1);
        ops.save(catalog2);
        System.out.println("Catalog ID : " + catalog1.getId());
                System.out.println("Catalog ID : " + catalog2.getId());
    }

    public static void saveDocuments() {

    }

    public static void removeDocument() {

    }

    public static void removeDocuments() {

    }
```

```
    public static void insertDocument() {

    }

    public static void insertDocuments() {

    }

    public static void documentExists() {

    }

    public static void findDocumentById() {

    }
    public static void findDocumentByView() {

    }

    public static void queryDocumentView() {

    }

    public static void updateDocument() {

    }

    public static void updateDocuments() {

    }
    public static void bucketCallback() {
        }

}
```

Next, run the App.java class. Right-click on App.java and select Run As ➤ Java Application as shown in Figure 3-7.

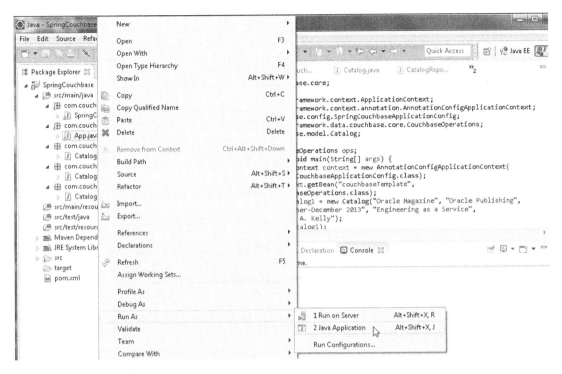

***Figure 3-7.*** *Running Maven Project App.java*

Two `Catalog` instances gets persisted to the "default" bucket. The Catalog ids of the persisted instance are output in the Console as shown in Figure 3-8.

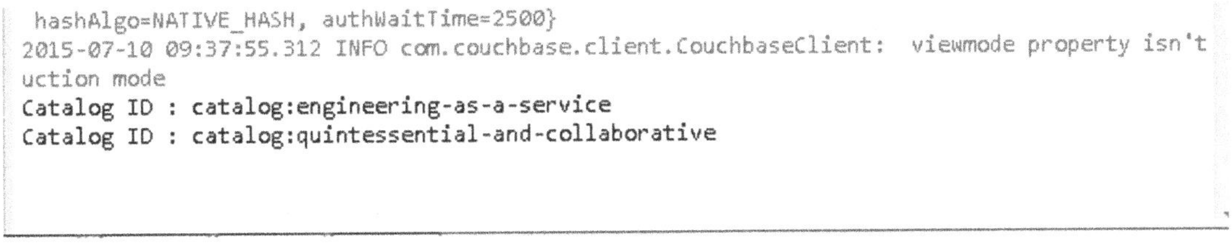

```
hashAlgo=NATIVE_HASH, authWaitTime=2500}
2015-07-10 09:37:55.312 INFO com.couchbase.client.CouchbaseClient:   viewmode property isn't
uction mode
Catalog ID : catalog:engineering-as-a-service
Catalog ID : catalog:quintessential-and-collaborative
```

***Figure 3-8.*** *Catalog id of Persisted Catalog Instance*

To list the JSON document objects added to Couchbase Server, click on the Documents button in the Couchbase Console. Two JSON document objects with ids `catalog:engineering-as-a-service` and `catalog:quintessential-and-collaborative` get listed. Click on Edit Document to display a document object as shown in Figure 3-9.

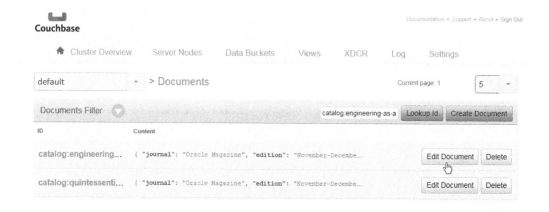

**Figure 3-9.** *Two Document Ids in Couchbase Server Console*

The JSON document object gets displayed in the Console as shown in Figure 3-10.

**Figure 3-10.** *Displaying a JSON document*

# Running Couchbase CRUD Operations

The CouchbaseOperations interface provides various methods for CRUD operations on a JSON document object stored in Couchbase Server. The different Couchbase operations are discussed next. First, the save operations.

## Save Ops

The CouchbaseOperations interface provides the methods discussed in Table 3-4 for the save operation.

***Table 3-4.*** *CouchbaseOperations interface Methods for saving documents*

| Method | Description |
|---|---|
| void save(Object objectToSave) | Saves an object. If the document with the same id already exists it is overridden. If a document with the id does not exist it is created. |
| void save(Collection<? extends Object> batchToSave) | Saves a collection of objects. If a document with the same id as in the collection already exists it is overridden. If a document with the id does not exist it is created. |

We created two Catalog instances and persisted them to Couchbase Server in the previous section. We invoked the save(Object objectToSave) method twice to save two documents. We could also use the save(Collection<? extends Object> batchToSave) method to save a batch of documents. To use the save(Collection<? extends Object> batchToSave) method create an ArrayList instance and add Catalog instances to it using ArrayList's add method. Before, running the application with the save(Collection<? extends Object> batchToSave) method delete the two documents persisted with the save(Object objectToSave) method as we shall add the same two Catalog instances with the save(Collection<? extends Object> batchToSave). Couchbase Server documents may be removed from the Couchbase Console by selecting Delete for a document.

```
public static void saveDocuments() {
ArrayList arrayList = new ArrayList();
    arrayList.add(catalog1);
    arrayList.add(catalog2);
    ops.save(arrayList);
}
```

When the App application, which is available with the downloads for this book (see the Source Code/ Downloads tab at www.apress.com/9781484214350), is run with only the saveDocuments() method invocation uncommented, two Catalog instances get saved in Couchbase Server. The two JSON document objects get listed in the Console again as shown in Figure 3-11.

***Figure 3-11.*** *The two JSON Documents Persisted to Couchbase using Batch Save*

## Remove Ops

The CouchbaseOperations interface provides the methods discussed in Table 3-5 for the remove operation.

**Table 3-5.** *CouchbaseOperations interface Methods for removing documents*

| Method | Description |
|---|---|
| void remove(Object object) | Removes an entity instance from the Couchbase document store. |
| void remove(Collection<? extends Object> batchToRemove) | Removes a Collection of entities from the Couchbase document store. |

As an example, remove the catalog1 instance, which was previously added using the remove(Object object) method.

```
public static void removeDocument() {
    ops.remove(catalog1);
}
```

When the App application is run with only the removeDocument() method invocation uncommented, the document object for the catalog1 entity instance gets removed from the Couchbase Server as shown by the single document in the Console in Figure 3-12.

**Figure 3-12.** *Listing only one Document as the other has been removed*

As an example remove a Collection of objects with the remove(Collection<? extends Object> batchToRemove) method. First, create an ArrayList instance for the documents to remove and subsequently invoke the remove(Collection<? extends Object> batchToRemove) method.

```
public static void removeDocuments() {
    arrayList.add(catalog1);
    arrayList.add(catalog2);
    ops.remove(arrayList);
}
```

Run the App application with only the saveDocuments() method invocation uncommented again before running the App with the removeDocuments() method invocation. All the document objects get removed from the Couchbase Server as shown in the Couchbase Console as shown in Figure 3-13.

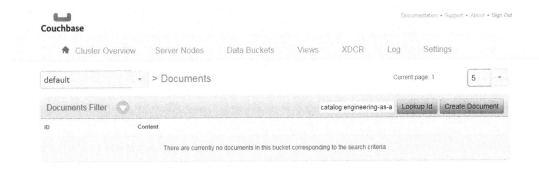

**Figure 3-13.** *Listing no Documents as all have been removed*

## Insert Ops

The CouchbaseOperations interface provides the overloaded insert methods discussed in Table 3-6 for the insert operation.

**Table 3-6.** *CouchbaseOperations interface Methods for adding documents*

| Method | Description |
| --- | --- |
| void insert(Object objectToSave) | Inserts an object to the Couchbase Server. The difference between the insert and save method is that the insert method does not replace an object if an object by the same Id already exists in the Couchbase Server data bucket. An error is not generated either if a document with the same id already exists in the data bucket. |
| void insert(Collection<? extends Object> batchToSave) | Inserts a Collection of objects in the Couchbase Server. The difference between the insert and save method is that the insert method does not replace an object if an object by the same Id already exists in the Couchbase Server data bucket. An error is not generated either if a document with the same id already exists in the data bucket. |

As an example of using the insert(Object objectToSave) method, add the catalog1 instance. If the catalog1 instance was previously added with the save method, remove the previously added document object.

```
public static void insertDocument() {
    ops.insert(catalog1);
    }
```

As an example of using the insert(Collection<? extends Object> batchToSave) method, add an ArrayList instance arrayList. If the arrayList was previously added with the save method, remove the previously added document objects.

```
public static void insertDocuments() {
        arrayList.add(catalog1);
        arrayList.add(catalog2);
        ops.insert(arrayList);
    }
```

## Exists Method

The exists(String id) method may be used to find out if a document with a given Id exists in the Couchbase Server. For example, find if documents with the ids catalog:engineering-as-a-service and catalog:quintessential-and-collaborative exist.

```
public static void documentExists() {
        System.out.println("catalog:engineering-as-a-service ID exists: "
            + ops.exists("catalog:engineering-as-a-service"));
        System.out.println("catalog:quintessential-and-collaborative ID exists: "
            + ops.exists("catalog:quintessential-and-collaborative"));
    }
```

The output in Eclipse Console indicates that the documents exist as shown in Figure 3-14.

```
2015-07-10 10:19:13.750 INFO com.couchbase.client.CouchbaseClient:  viewmode property isn't
uction mode
catalog:engineering-as-a-service ID exists: true
catalog:quintessential-and-collaborative ID exists: true
```

***Figure 3-14.*** *Output from invoking the exists method*

## Find Ops

The CouchbaseOperations interface provides the methods discussed in Table 3-7 for the find operation.

***Table 3-7.*** *CouchbaseOperations interface Methods for finding documents*

| Method | Description |
|---|---|
| <T> T findById(String id, Class<T> entityClass) | Finds an object by the given entity and maps it to the specified entity class. |
| <T> List<T> findByView(String design, String view, com.couchbase.client. protocol.views.Query query, Class<T> entityClass) | Queries a View using the specified design document and Query object for a list of documents and maps the documents to the specified entity class. |

As an example, find a document with id `catalog:engineering-as-a-service` and map the document to the `Catalog` type. Output the properties of the `Catalog` instance using the get methods.

```
public static void findDocumentById() {
        Catalog catalog = ops.findById("catalog:engineering-as-a-service",Catalog.class);
        System.out.println("Journal : " + catalog.getJournal());
        System.out.println("Publisher : " + catalog.getPublisher());
        System.out.println("Edition : " + catalog.getEdition());
        System.out.println("Title : " + catalog.getTitle());
        System.out.println("Author : " + catalog.getAuthor());
}
```

When the App application is run with only the findDocumentById() method invocation uncommented, the output is shown in the Eclipse Console in Figure 3-15.

```
2015-07-10 10:20:50.208 INFO com.couchbase.client.CouchbaseClient:  viewmode property isn't
uction mode
Journal : Oracle Magazine
Publisher : Oracle Publishing
Edition : November-December 2013
Title : Engineering as a Service
Author : David A. Kelly
```

***Figure 3-15.*** *Output from findById method*

For using the `findByView` method we need to create a design document and a view in the Couchbase Server. To add a design document we need to create a `CouchbaseClient` instance using a `List` of `URI`s and the "default" bucket. Create a `DesignDocument` object using the `DesignDocument(String)` constructor in the findDocumentByView() class method of the App application.

```
DesignDocument designDoc = new DesignDocument("JSONDocument");
```

Create a `ViewDesign` object using a view name and a map function. Add the `ViewDesign` instance to the `DesignDocument` instance using the `getViews()` method and subsequently the add method.

```
ViewDesign viewDesign = new ViewDesign(viewName, mapFunction);
designDoc.getViews().add(viewDesign);
```

Store a `DesignDocument` in the Couchbase cluster using the `asyncCreateDesignDoc(DesignDocument doc)` method of `com.couchbase.client.CouchbaseClient`. Get access to the view contained in the design document using the `getView(java.lang.String designDocumentName,java.lang.String viewName)` method in `CouchbaseClient`.

```
    HttpFuture<java.lang.Boolean> httpFuture = couchbaseClient
        .asyncCreateDesignDoc(designDoc);
View view = couchbaseClient.getView("JSONDocument", "by_name");
```

What the view does is to take the structured JSON document stored in Couchbase Server and extracts and indexes its fields, allowing you to query the stored data. The Query class is used to create custom view queries. To include full documents in the result invoke the `setIncludeDocs(boolean include)` method of the Query class. Optionally limit the number of documents using the `setLimit(int limit)` method of Query class. To disallow results from a stale view to be used, invoke the setStale(Stale stale) method from Query using Stale.FALSE as argument.

```
Query query = new Query();
query.setIncludeDocs(true).setLimit(20);
query.setStale(Stale.FALSE);
```

Invoke the findByView method using the design document, view name, Query instance, and the entity class as arguments.

```
List<Catalog> catalogList = ops.findByView("JSONDocument",viewName, query, Catalog.class);
```

Finally, obtain an Iterator from the List and use the Iterator to iterate over the List and output the Catalog entity instance fields. The finished method is shown below:

```
public static void findDocumentByView(){
   List<URI> uris = new LinkedList<URI>();
   uris.add(URI.create("http://127.0.0.1:8091/pools"));
   CouchbaseClient couchbaseClient;
     try {
         couchbaseClient = new CouchbaseClient(uris, "default", "");
      DesignDocument designDoc = new DesignDocument("JSONDocument");
      String viewName = "by_name";
      String mapFunction = "  function(doc,meta) {\n"
           + "  if (meta.type == 'json') {\n"
           + "  emit(doc.name, [doc.journal,doc.publisher,doc.edition,doc.title,doc.
           author]);\n"
           + "  }\n" + "}";
      ViewDesign viewDesign = new ViewDesign(viewName, mapFunction);
      designDoc.getViews().add(viewDesign);
         HttpFuture<java.lang.Boolean> httpFuture = couchbaseClient
              .asyncCreateDesignDoc(designDoc);
      View view = couchbaseClient.getView("JSONDocument", "by_name");
      Query query = new Query();
      query.setIncludeDocs(true).setLimit(20);
      query.setStale(Stale.FALSE);
      List<Catalog> catalogList = ops.findByView("JSONDocument",
           viewName, query, Catalog.class);
      Iterator<Catalog> iter = catalogList.iterator();
      while (iter.hasNext()) {
         Catalog catalog = iter.next();
         System.out.println("Journal : " + catalog.getJournal());
         System.out.println("Publisher : " + catalog.getPublisher());
         System.out.println("Edition : " + catalog.getEdition());
         System.out.println("Title : " + catalog.getTitle());
         System.out.println("Author : " + catalog.getAuthor());
      }
   } catch (UnsupportedEncodingException e) {
```

```
            e.printStackTrace();
    } catch (IOException e) {

            e.printStackTrace();
    }
}
```

Invoke only the `findDocumentByView` method in the `main` method and run the App application. The result is shown in Eclipse Console in Figure 3-16.

```
2015-07-10 10:24:17.131 INFO com.couchbase.client.CouchbaseClient:  Creating Design Document:JSONDocument
Journal : Oracle Magazine
Publisher : Oracle Publishing
Edition : November-December 2013
Title : Engineering as a Service
Author : David A. Kelly
Journal : Oracle Magazine
Publisher : Oracle Publishing
Edition : November-December 2013
Title : Quintessential and Collaborative
Author : Tom Haunert
```

***Figure 3-16.*** *Output from findByView method*

A by_name View also gets created in a `JSONDocument` design document as shown in Figure 3-17.

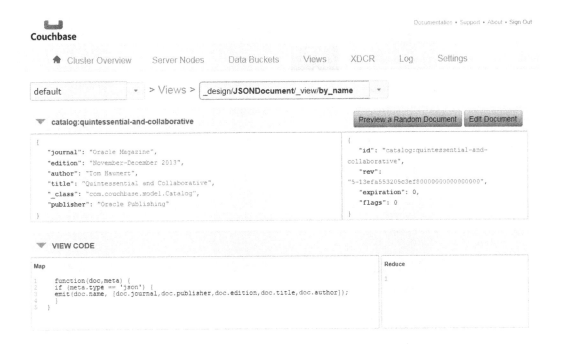

***Figure 3-17.*** *The by_name View*

## Query View

A View can be queried with direct access to the ViewResponse object using the queryView(String design,String view, com.couchbase.client.protocol.views.Query query) method, which returns a ViewResponse object. The design argument is the design document name, the view argument is the view name, and the query argument is the view query. In the queryDocumentView() class method in App application create a ViewResponse instance.

```
com.couchbase.client.protocol.views.ViewResponse viewResponse =
ops.queryView("JSONDocument", viewName, query);
```

Iterate over the ViewResponse object to generate ViewRows and output the Id, key, and value in each view row using the corresponding get methods.

```
for (ViewRow row : viewResponse) {
   System.out.println("Id " + row.getId());
   System.out.println("Key " + row.getKey());
   System.out.println("Value " + row.getValue());
}
```

Invoke the findDocumentByView() and queryDocumentView() in sequence as the viewName and query are set in the findDocumentByView() method.

```
findDocumentByView();
queryDocumentView();
```

When the App application is run. the output from the ViewResponse obtained with the queryView method is shown in Eclipse Console in Figure 3-18.

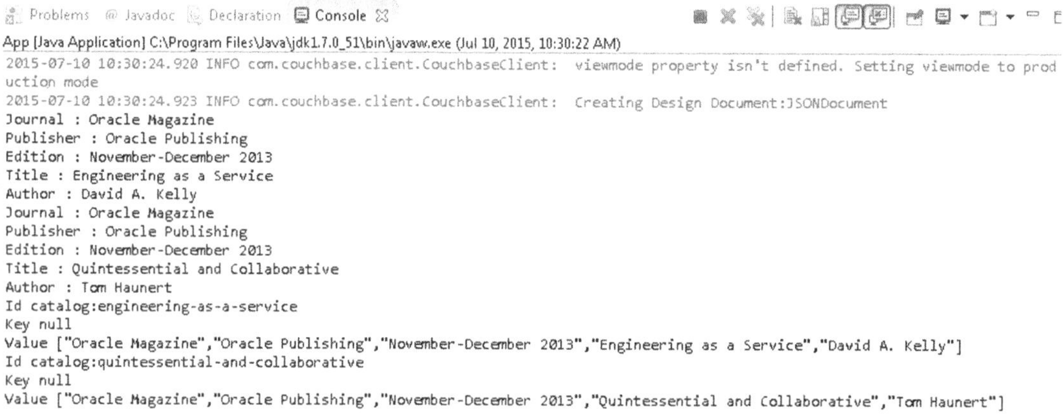

**Figure 3-18.** *Output from queryView method*

## Update Ops

The CouchbaseOperations interface provides the methods discussed in Table 3-8 for the update operation.

**Table 3-8.** *CouchbaseOperations interface Methods for updating documents*

| Method | Description |
| --- | --- |
| `void update(Object objectToSave)` | Updates the given object. If the update is invoked on a document that is not in the Couchbase datastore, a new document object is not created, and the method invocation is effectively ignored. |
| `void update(Collection<? extends Object> batchToSave)` | Updates the given Collection of objects. If one of the documents does not exist in the Couchbase datastore, a new document object is not created, and the method invocation is effectively ignored. |

As an example, update the document with title *Engineering as a Service*, which is stored as id `catalog:engineering-as-a-service`. Specify different values for the edition and author fields in the updated `Catalog` instance with title *Engineering as a Service*. Update the document object using the `update(Object objectToSave)` method.

```
public static void updateDocument() {
        catalog1 = new Catalog("Oracle Magazine", "Oracle Publishing",
                "11/12 2013", "Engineering as a Service", "Kelly, David A.");
        ops.update(catalog1);
}
```

When the preceding example is run, the document object gets updated as shown in the Console in Figure 3-19.

**Figure 3-19.** *Updating a Document with the update method*

As an example of using the `update(Collection<? extends Object> batchToSave)` method, create an `ArrayList` instance for `Catalog` ids to be updated. If the preceding examples have been run Couchbase Server should have two documents stored. If not, first save the entity instances to be updated using one of the `save` methods. Subsequently, invoke the `update(Collection<? extends Object> batchToSave)` method on the `ArrayList` instance.

```
public static void updateDocuments() {
        Catalog catalog1 = new Catalog("Oracle Magazine", "Oracle Publishing",
                "November December 2013", "Engineering as a Service", "David Kelly");
```

```
        Catalog catalog2 = new Catalog("Oracle Magazine", "Oracle Publishing",
                "11/12 2013", "Quintessential and Collaborative",
                "Haunert, Tom");
        arrayList = new ArrayList();
        arrayList.add(catalog1);
        arrayList.add(catalog2);
        ops.update(arrayList);

    }
```

When the preceding example is run the document objects in the Couchbase datastore get updated.

## Bucket Callback

The org.springframework.data.couchbase.core.BucketCallback<T> interface defines a callback, which is wrapped and executed on a bucket. The callback may be used to add/update/delete data in the connected bucket. The execute(BucketCallback<T> action) method in CouchbaseOperations may be used to execute a BucketCallback. The doInBucket() method is the only method in the BucketCallback interface. As an example, invoke the execute(BucketCallback<T> action) method and use an anonymous class as an argument to the method. Output the fields of the Catalog instance added to the document store in the doInBucket() method.

```
Catalog catalog = ops.execute(new BucketCallback<Catalog>() {
    public Catalog doInBucket() throws TimeoutException,
          ExecutionException, InterruptedException {
        Catalog catalog1 = new Catalog("Oracle_Magazine",
              "Oracle_Publishing", "11/12 2013",
              "Engineering_as_a_Service", "Kelly, David");
        ops.save(catalog1);
        return catalog1;
    }
});
System.out.println("Journal : " + catalog.getJournal());
System.out.println("Publisher : " + catalog.getPublisher());
System.out.println("Edition : " + catalog.getEdition());
System.out.println("Title : " + catalog.getTitle());
System.out.println("Author : " + catalog.getAuthor());
```

The output from running the preceding example is shown in Eclipse Console in Figure 3-20.

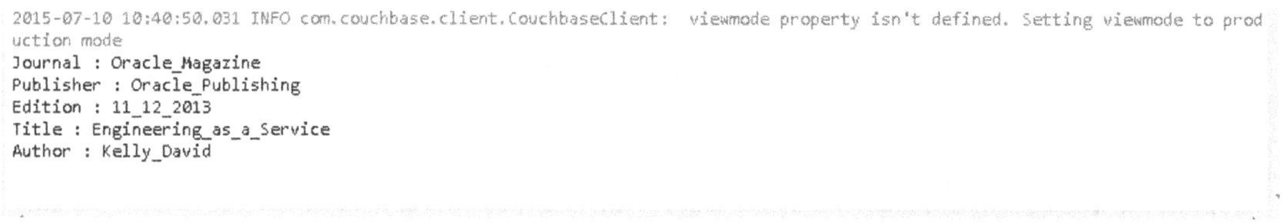

```
2015-07-10 10:40:50.031 INFO com.couchbase.client.CouchbaseClient:  viewmode property isn't defined. Setting viewmode to prod
uction mode
Journal : Oracle_Magazine
Publisher : Oracle_Publishing
Edition : 11_12_2013
Title : Engineering_as_a_Service
Author : Kelly_David
```

***Figure 3-20.*** *Output from using the BucketCallback Interface*

The App.java application used to run CRUD operations using the CouchbaseOperations is listed; this is the result of all the code we have looked at in this chapter so far.

```java
package com.couchbase.core;

import java.io.IOException;
import java.io.UnsupportedEncodingException;
import java.net.URI;
import java.util.ArrayList;
import java.util.Iterator;
import java.util.LinkedList;
import java.util.List;
import java.util.concurrent.ExecutionException;
import java.util.concurrent.TimeoutException;
import org.springframework.context.ApplicationContext;
import org.springframework.context.annotation.AnnotationConfigApplicationContext;
import org.springframework.data.couchbase.core.BucketCallback;
import org.springframework.data.couchbase.core.CouchbaseOperations;
import com.couchbase.model.Catalog;
import com.couchbase.client.CouchbaseClient;
import com.couchbase.client.internal.HttpFuture;
import com.couchbase.client.protocol.views.DesignDocument;
import com.couchbase.client.protocol.views.Query;
import com.couchbase.client.protocol.views.Stale;
import com.couchbase.client.protocol.views.View;
import com.couchbase.client.protocol.views.ViewDesign;
import com.couchbase.client.protocol.views.ViewRow;
import com.couchbase.config.SpringCouchbaseApplicationConfig;

public class App {
    static CouchbaseOperations ops;
    static Catalog catalog1;
    static Catalog catalog2;
    static ArrayList arrayList;
    static Query query;
    static String viewName;

    public static void main(String[] args) {
        ApplicationContext context = new AnnotationConfigApplicationContext(
                SpringCouchbaseApplicationConfig.class);
        ops = context.getBean("couchbaseTemplate", CouchbaseOperations.class);
        catalog1 = new Catalog("Oracle Magazine", "Oracle Publishing",
                "November-December 2013", "Engineering as a Service",
                "David A. Kelly");
        ops.save(catalog1);
        catalog2 = new Catalog("Oracle Magazine", "Oracle Publishing",
                "November-December 2013", "Quintessential and Collaborative",
                "Tom Haunert");
        arrayList = new ArrayList();
        // saveDocument();
        // saveDocuments();
        // removeDocument();
```

```java
        // removeDocuments();
        // insertDocument();
        // insertDocuments();
        // documentExists();
        // findDocumentById();
        // findDocumentByView();
        // queryDocumentView();
        // updateDocument();
        // updateDocuments();
        bucketCallback();
    }

    public static void saveDocument() {
        ops.save(catalog1);
        ops.save(catalog2);
        System.out.println("Catalog ID : " + catalog1.getId());

    public static void saveDocuments() {
        arrayList.add(catalog1);
        arrayList.add(catalog2);
        ops.save(arrayList);
    }

    public static void removeDocument() {
        ops.remove(catalog1);
    }

    public static void removeDocuments() {
        arrayList.add(catalog1);
        arrayList.add(catalog2);
        ops.remove(arrayList);
    }

    public static void insertDocument() {
        ops.insert(catalog1);
    }

    public static void insertDocuments() {
        arrayList.add(catalog1);
        arrayList.add(catalog2);
        ops.insert(arrayList);
    }

    public static void documentExists() {
        System.out.println("catalog:engineering-as-a-service ID exists: "
                + ops.exists("catalog:engineering-as-a-service"));

        System.out
                .println("catalog:quintessential-and-collaborative ID exists: "
                        + ops.exists("catalog:quintessential-and-collaborative"));
    }
```

```java
public static void findDocumentById() {
    Catalog catalog = ops.findById("catalog:engineering-as-a-service",
            Catalog.class);
    System.out.println("Journal : " + catalog.getJournal());
    System.out.println("Publisher : " + catalog.getPublisher());
    System.out.println("Edition : " + catalog.getEdition());
    System.out.println("Title : " + catalog.getTitle());
    System.out.println("Author : " + catalog.getAuthor());
}

public static void findDocumentByView() {
    List<URI> uris = new LinkedList<URI>();
    uris.add(URI.create("http://127.0.0.1:8091/pools"));
    CouchbaseClient couchbaseClient;
    try {
        couchbaseClient = new CouchbaseClient(uris, "default", "");
        DesignDocument designDoc = new DesignDocument("JSONDocument");
        viewName = "by_name";
        String mapFunction = "  function(doc,meta) {\n"
                + "  if (meta.type == 'json') {\n"
                + "  emit(doc.name, [doc.journal,doc.publisher,doc.edition,doc.
                title,doc.author]);\n"
                + "  }\n" + "}";
        ViewDesign viewDesign = new ViewDesign(viewName, mapFunction);
        designDoc.getViews().add(viewDesign);
        HttpFuture<java.lang.Boolean> httpFuture = couchbaseClient
                .asyncCreateDesignDoc(designDoc);
        View view = couchbaseClient.getView("JSONDocument", "by_name");
        query = new Query();
        query.setIncludeDocs(true).setLimit(20);
        query.setStale(Stale.FALSE);
        List<Catalog> catalogList = ops.findByView("JSONDocument",
                viewName, query, Catalog.class);
        Iterator<Catalog> iter = catalogList.iterator();
        while (iter.hasNext()) {
            Catalog catalog = iter.next();
            System.out.println("Journal : " + catalog.getJournal());
            System.out.println("Publisher : " + catalog.getPublisher());
            System.out.println("Edition : " + catalog.getEdition());
            System.out.println("Title : " + catalog.getTitle());
            System.out.println("Author : " + catalog.getAuthor());
        }
    } catch (UnsupportedEncodingException e) {
        e.printStackTrace();
    } catch (IOException e) {
        e.printStackTrace();
    }

}
```

```
    public static void queryDocumentView() {
        com.couchbase.client.protocol.views.ViewResponse viewResponse = ops
                .queryView("JSONDocument", "by_name", query);
        for (ViewRow row : viewResponse) {
            System.out.println("Id " + row.getId());
            System.out.println("Key " + row.getKey());
            System.out.println("Value " + row.getValue());
        }
    }

    public static void updateDocument() {
        catalog1 = new Catalog("Oracle Magazine", "Oracle Publishing",
                "11/12 2013", "Engineering as a Service", "Kelly, David A.");
        ops.update(catalog1);
    }

    public static void updateDocuments() {
        Catalog catalog1 = new Catalog("Oracle Magazine", "Oracle Publishing",
                "November December 2013", "Engineering as a Service",
                "David Kelly");
        Catalog catalog2 = new Catalog("Oracle Magazine", "Oracle Publishing",
                "11/12 2013", "Quintessential and Collaborative",
                "Haunert, Tom");
        arrayList = new ArrayList();
        arrayList.add(catalog1);
        arrayList.add(catalog2);
        ops.update(arrayList);

    }

    public static void bucketCallback() {
        Catalog catalog = ops.execute(new BucketCallback<Catalog>() {
            public Catalog doInBucket() throws TimeoutException,
                    ExecutionException, InterruptedException {
                Catalog catalog1 = new Catalog("Oracle_Magazine",
                        "Oracle_Publishing", "11_12_2013",
                        "Engineering_as_a_Service", "Kelly_David");
                ops.save(catalog1);
                return catalog1;
            }
        });
        System.out.println("Journal : " + catalog.getJournal());
        System.out.println("Publisher : " + catalog.getPublisher());
        System.out.println("Edition : " + catalog.getEdition());
        System.out.println("Title : " + catalog.getTitle());
        System.out.println("Author : " + catalog.getAuthor());
    }
}
```

# Using Spring Data Repositories with Couchbase

Spring Data Repositories are an abstraction that implement a data access layer over the underlying datastore. Spring Data Repositories reduce the boilerplate code required to access a datastore. Spring Data Repositories may be used with Couchbase datastore. The automatic implementation of Repository interfaces including using for custom finder methods is one of the most important features of Couchbase. To enable the Spring Data Repositories infrastructure for Couchbase, annotate the JavaConfig class with @EnableCouchbaseRepositories. Annotate the JavaConfig SpringCouchbaseApplicationConfig class with @EnableCouchbaseRepositories("com.couchbase.repositories"). com.couchbase.repositories is the package to search for repositories.

```
@Configuration
@EnableCouchbaseRepositories("com.couchbase.repositories")
public class SpringCouchbaseApplicationConfig extends AbstractCouchbaseConfiguration {
}
```

The central repository marker interface is org.springframework.data.repository.Repository, and it provides CRUD access on top of the entities. The entity class Catalog is defined earlier in the chapter. The generic interface for CRUD operations on a repository is org.springframework.data.repository .CrudRepository. The Couchbase Server specific repository interface is org.springframework.data .couchbase.repository.CouchbaseRepository<T,ID extends Serializable>. The interface is parameterized over the domain type, which would be Catalog for the example and ID type, which is String in the example. The ID extends the java.io.Serializable interface to be able to serialize the ID in the Couchbase Server. Create an interface, CatalogRepository, which extends the parameterized type CouchbaseRepository<Catalog, String>. The CatalogRepository represents the Couchbase-specific repository interface to store entities of type Catalog and with Id of type String in Couchbase Server.

```
package com.couchbase.repositories;
import org.springframework.data.couchbase.repository.CouchbaseRepository;
import com.couchbase.model.Catalog;
public interface CatalogRepository extends CouchbaseRepository<Catalog, String>{
}
```

Create a service class CatalogService and in the class create a repository instance from the context as follows.

```
ApplicationContext context = new AnnotationConfigApplicationContext
(SpringCouchbaseApplicationConfig.class);
CatalogRepository repository = context.getBean(CatalogRepository.class);
```

Subsequently, we shall perform CRUD operations on the Couchbase document store using the CatalogRepository instance. But, first we need to create an "all" View for the data bucket in which the documents are stored. An "all" view is required for some of the methods such as findAll(), deleteAll() and count(), which do not make use of a document id.

## Creating the all View

A view extracts and indexes the fields of a document store for subsequent querying. To do so, you need to specify a List of URIs for the Couchbase Server; in the code below we specify only the http://127.0.0.1:8091/pools URI. Create a CouchbaseClient instance using the List of URIs and the "default" bucket.

```
List<URI> uris = new LinkedList<URI>();
uris.add(URI.create("http://127.0.0.1:8091/pools"));
CouchbaseClient couchbaseClient;
try {
   couchbaseClient = new CouchbaseClient(uris, "default", "");
```

Create a DesignDocument "catalog." Specify view name as "all" and specify map and reduce functions to create a ViewDesign instance from.

```
DesignDocument designDoc = new DesignDocument("catalog");
String viewName = "all";
 String mapFunction = "  function(doc,meta) {"
       + "  if (meta.type == 'json') {"
       + "  emit(doc.name, [doc.journal,doc.publisher,doc.edition,doc.title,doc.author]);"
       + "  }" + "}";
String reduceFunction = "function(key, values, rereduce) {"
       + "if (rereduce) {" + "var result = 0;"
       + "for (var i = 0; i < values.length; i++) {"
       + "result += values[i];" + "}" + "return result;"
       + "  } else {" + " return values.length;" + " }" + "}";
```

Add the ViewDesign instance to the DesignDocument instance. Create the DesignDocument instance in the Couchbase Server using the CouchbaseClient instance.

```
ViewDesign viewDesign = new ViewDesign(viewName, mapFunction,
       reduceFunction);
designDoc.getViews().add(viewDesign);
HttpFuture<java.lang.Boolean> httpFuture = couchbaseClient
       .asyncCreateDesignDoc(designDoc);
```

The partial CatalogService class in which the "all" view is created is listed below.

```
package service;

import java.io.IOException;
import java.net.URI;
import java.util.ArrayList;
import java.util.Iterator;
import java.util.LinkedList;
import java.util.List;
import org.springframework.beans.factory.annotation.Autowired;
import org.springframework.context.ApplicationContext;
import org.springframework.context.annotation.AnnotationConfigApplicationContext;
import com.couchbase.client.CouchbaseClient;
import com.couchbase.client.internal.HttpFuture;
import com.couchbase.client.protocol.views.DesignDocument;
import com.couchbase.client.protocol.views.ViewDesign;
import com.couchbase.config.SpringCouchbaseApplicationConfig;
import com.couchbase.model.Catalog;
import com.couchbase.repositories.CatalogRepository;
```

```
public class CatalogService {
// @Autowired
// public static CatalogRepository repository;
   public static void main(String[] args) {
       /**ApplicationContext context = new AnnotationConfigApplicationContext(
           SpringCouchbaseApplicationConfig.class);
       repository = context.getBean(CatalogRepository.class);*/
       List<URI> uris = new LinkedList<URI>();
       uris.add(URI.create("http://127.0.0.1:8091/pools"));
       CouchbaseClient couchbaseClient;
       try {
          couchbaseClient = new CouchbaseClient(uris, "default", "");
          DesignDocument designDoc = new DesignDocument("catalog");
          String viewName = "all";
          String mapFunction = "  function(doc,meta) {"
               + "  if (meta.type == 'json') {"
               + "  emit(doc.name, [doc.journal,doc.publisher,doc.edition,doc.title,
               doc.author]);"
               + "  }" + "}";
          String reduceFunction = "function(key, values, rereduce) {"
               + "if (rereduce) {" + "var result = 0;"
               + "for (var i = 0; i < values.length; i++) {"
               + "result += values[i];" + "}" + "return result;"
               + "  } else {" + " return values.length;" + " }" + "}";
          ViewDesign viewDesign = new ViewDesign(viewName, mapFunction,
               reduceFunction);
          designDoc.getViews().add(viewDesign);
          HttpFuture<java.lang.Boolean> httpFuture = couchbaseClient
               .asyncCreateDesignDoc(designDoc);
       } catch (IOException e) {
          e.printStackTrace();
       }
   }
}
```

Right-click on the CatalogService class in Package Explorer and select Run As ➤ Java Application.

The "all" view gets added on the "default" bucket. In the Console click on Views to list the "all" view. Click on the "all" view to list the map and reduce functions and the Key/Value result for the view as shown in Figure 3-21.

**Figure 3-21.** *Listing the all View*

Next, we shall perform CRUD operations on the document stored in the Couchbase Server "default" bucket using the CatalogRepository instance. Add the class methods listed in Table 3-9 to the CatalogService class. Invoke the methods in the main method to perform CRUD operations using Couchbase Repository.

**Table 3-9.** *Class Methods*

| Method | Description |
|---|---|
| countDocuments() | Counts number of documents. |
| findAllDocuments() | Finds all documents. |
| findOneDocument() | Finds a single document. |
| findDocumentExists() | Finds if a document exists. |
| saveDocument() | Saves a document. |
| saveDocuments() | Saves a batch of documents. |
| deleteDocument() | Deletes a single document. |
| deleteDocuments() | Deletes a batch of documents. |

# Document Count

The CouchbaseRepository<T,ID extends Serializable> interface, which extends the CrudRepository<T,ID> interface provides CRUD operation methods. The CouchbaseRepository interface also includes a count() method, which returns the number of entities stored in a data bucket. The count() method makes use of the "all" view. To be able to count entities, first create some entities using the CouchbaseOperations instance as discussed earlier in the Running Couchbase CRUD Operations section.

Invoke the count() method using the CatalogRepository instance and output the long value returned.

```java
public static void countDocuments() {
    long count = repository.count();
    System.out.println("Number of catalogs: " + count);
}
```

The output in the Eclipse Console in Figure 3-22 shows that the number of document entities stored is 3.

```
Problems  Javadoc  Declaration  Console
CatalogService [Java Application] C:\Program Files\Java\jdk1.7.0_51\bin\javaw.exe (Jul 10, 2015, 11:07:26 AM)
2015-07-10 11:07:30.352 INFO com.couchbase.client.vbucket.provider.BucketConfigurationProvider:  Could bootstrap through carr
ier publication.
2015-07-10 11:07:30.356 INFO com.couchbase.client.CouchbaseConnection:  Added {QA sa=/127.0.0.1:11210, #Rops=0, #Wops=0, #iq=
0, topRop=null, topWop=null, toWrite=0, interested=0} to connect queue
2015-07-10 11:07:30.357 INFO com.couchbase.client.CouchbaseClient:  CouchbaseConnectionFactory{bucket='default', nodes=[http:
//127.0.0.1:8091/pools], order=RANDOM, opTimeout=2500, opQueue=16384, opQueueBlockTime=10000, obsPollInt=10, obsPollMax=500,
obsTimeout=5000, viewConns=10, viewTimeout=75000, viewWorkers=1, configCheck=10, reconnectInt=1100, failureMode=Redistribute,
hashAlgo=NATIVE_HASH, authWaitTime=2500}
2015-07-10 11:07:30.361 INFO com.couchbase.client.CouchbaseClient:  viewmode property isn't defined. Setting viewmode to prod
uction mode
Number of catalogs: 3
```

***Figure 3-22.*** *Counting Documents*

# Finding Entities from the Repository

The CouchbaseRepository interface provides the methods discussed in Table 3-10 for finding documents from the repository.

***Table 3-10.*** *CouchbaseRepository interface Methods for Finding documents*

| Method | Description |
|---|---|
| T findOne(ID id) | Returns the entity instance for the specified ID. |
| Iterable<T> findAll() | Returns all instances of the entity type specified in the repository. |
| Iterable<T> findAll(Iterable<ID> ids) | Returns all instances of an entity type for the given IDs. |

As an example, find all instances from the CatalogRepository instance repository using the findAll() method.

```java
Iterable<Catalog> iterable = repository.findAll();
```

The findAll() method returns an Iterable from which it obtains an Iterator using the iterator() method.

```java
Iterator<Catalog> iter = iterable.iterator();
```

Iterate over the entity instances using the Iterator and output the fields for each entity instance.

```
while (iter.hasNext()) {
   Catalog catalog = iter.next();
   System.out.println("Journal: " + catalog.getJournal());
   System.out.println("Publisher: " + catalog.getPublisher());
   System.out.println("Edition: " + catalog.getEdition());
   System.out.println("Title: " + catalog.getTitle());
   System.out.println("Author: " + catalog.getAuthor());
}
```

When the preceding example is run the field values for the two entity instances are output in the Eclipse Console as shown in Figure 3-23.

```
Journal: Oracle Magazine
Publisher: Oracle Publishing
Edition: November-December 2013
Title: Engineering as a Service
Author: David A. Kelly
Journal: Oracle_Magazine
Publisher: Oracle_Publishing
Edition: 11_12_2013
Title: Engineering_as_a_Service
Author: Kelly_David
Journal: Oracle Magazine
Publisher: Oracle Publishing
Edition: 11/12 2013
Title: Quintessential and Collaborative
Author: Haunert, Tom
```

***Figure 3-23.*** *Field Values for the Two entity Instances*

As an example of using the findOne method, find the entity instance with Id catalog:engineering-as-a-service. Subsequently output the field values for the entity instance.

```
Catalog catalog = repository.findOne("catalog:engineering-as-a-service");
        System.out.println("Journal: " + catalog.getJournal());
        System.out.println("Publisher: " + catalog.getPublisher());
        System.out.println("Edition: " + catalog.getEdition());
        System.out.println("Title: " + catalog.getTitle());
        System.out.println("Author: " + catalog.getAuthor());
```

The field values for the entity instance are output in the Eclipse Console as shown in Figure 3-24.

```
Problems  Javadoc  Declaration  Console
CatalogService [Java Application] C:\Program Files\Java\jdk1.7.0_51\bin\javaw.exe (Jul 10, 2015, 11:11:25 AM)
0, topRop=null, topWop=null, toWrite=0, interested=0) to connect queue
2015-07-10 11:11:29.918 INFO com.couchbase.client.CouchbaseClient:  CouchbaseConnectionFactory{bucket='default', nodes=[http:
//127.0.0.1:8091/pools], order=RANDOM, opTimeout=2500, opQueue=16384, opQueueBlockTime=10000, obsPollInt=10, obsPollMax=500,
obsTimeout=5000, viewConns=10, viewTimeout=75000, viewWorkers=1, configCheck=10, reconnectInt=1100, failureMode=Redistribute,
 hashAlgo=NATIVE_HASH, authWaitTime=2500}
2015-07-10 11:11:29.922 INFO com.couchbase.client.CouchbaseClient:  viewmode property isn't defined. Setting viewmode to prod
uction mode
Journal: Oracle Magazine
Publisher: Oracle Publishing
Edition: November-December 2013
Title: Engineering as a Service
Author: David A. Kelly
```

***Figure 3-24.*** *Output from findOne method*

## Finding if an Entity Exists

The CouchbaseRepository interface provides the exists(ID id) method to determine if an entity instance by the given id exists. As an example find if the catalog:quintessential-and-collaborative Id exists in the Couchbase datastore.

```
boolean bool = repository.exists("catalog:quintessential-and-collaborative");
System.out .println("Catalog with Id catalog:quintessential-and-collaborative exists: " +bool);
```

The output from invoking the exists() method indicates that the catalog:quintessential-and-collaborative Id exists in the datastore as shown in Figure 3-25.

```
2015-07-10 11:13:07.576 INFO com.couchbase.client.CouchbaseClient:  viewmode property isn't defined. Setting viewmode to prod
uction mode
Catalog with Id catalog:quintessential-and-collaborative exists: true
```

*Figure 3-25.* *Output from exists() method*

## Saving Entities

The CouchbaseRepository interface provides the methods discussed in Table 3-11 for saving documents using the repository. As discussed in Table 3-8 the save methods may also be used for updating an entity.

*Table 3-11.* *CouchbaseRepository interface Methods for saving documents*

| Method | Description |
| --- | --- |
| <S extends T> S save(S entity) | Saves a given entity. If the entity is already in the server, overwrites the entity. Returns the saved entity instance. |
| <S extends T> Iterable<S> save(Iterable<S> entities) | Saves a given collection of entities. If an entity is already in the server, overwrites the entity. Returns the saved entity instances. |

As an example create and save an entity instance using the save(S entity) method.

```
public static void saveDocument() {
    Catalog catalog = new Catalog("Oracle Magazine", "Oracle Publishing",
    "11/12 2013", "Engineering as a Service", "Kelly, David");
    repository.save(catalog);
}
```

When the CatalogService application is run with only the saveDocument() class method invocation in the main method an entity instance with Id generated from the title *Engineering as a Service* as specified in the entity class constructor gets saved as shown in the Console as shown in Figure 3-26.

*Figure 3-26.* *Saving a document with the save method*

As an example of using the save(Iterable<S> entities) method create an ArrayList of entity instances.

```
ArrayList arrayList = new ArrayList();
Catalog catalog1 = new Catalog("Oracle_Magazine", "Oracle_Publishing",
"November-December_2013", "EngineeringasaService",
"David_A._Kelly");
Catalog catalog2 = new Catalog("Oracle_Magazine", "Oracle_Publishing",
"11/12_2013", "Engineering_as_a_Service", "Kelly, David");
arrayList.add(catalog1);
arrayList.add(catalog2);
```

Invoke the save(Iterable<S> entities) method on the ArrayList instance.

```
repository.save(arrayList);
```

When the preceding example is run by invoking the saveDocuments() class method in CatalogService application the collection of entity instances in the ArrayList get added to the Couchbase Server as shown in Figure 3-27.

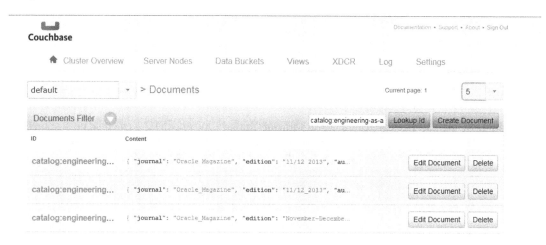

*Figure 3-27.* *Saving a Collection of documents*

# Deleting Entities

The CouchbaseRepository interface provides the methods discussed in Table 3-12 for deleting documents using the repository.

***Table 3-12.*** *CouchbaseRepository interface Methods for deleting documents*

| Method | Description |
|---|---|
| void delete(ID id) | Deletes the entity by the given ID managed by the repository. |
| void delete(T entity) | Deletes the specified entity managed by the repository. |
| void delete(Iterable<? extends T> entities) | Deletes the entities in the specified Iterable managed by the repository. |
| void deleteAll() | Deletes all entities managed by the repository. |

As an example of using the delete(ID id) method delete the entity with ID catalog:engineeringasaservice.

```
repository.delete("catalog:engineeringasaservice");
```

The entity with Id catalog:engineeringasaservice gets deleted as shown by the two remaining entities in the Console in Figure 3-28.

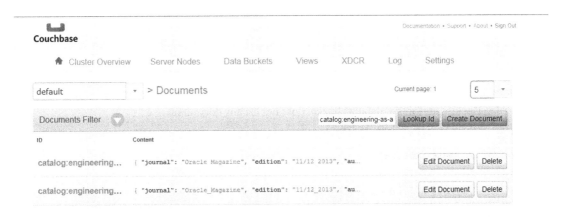

***Figure 3-28.*** *Listing two of the three documents*

As an example of using the delete(T entity) method delete the catalog2 entity instance, which was added previously.

```
Catalog catalog2 = new Catalog("Oracle_Magazine", "Oracle_Publishing",
"11/12_2013", "Engineering_as_a_Service", "Kelly, David");
repository.delete(catalog2);
```

The catalog2 entity gets deleted. Run the saveDocuments() again before the next delete to add a collection of entities as we deleted some in the preceding examples. As an example of using the delete(Iterable<? extends T> entities) method create an ArrayList of entity IDs and invoke the delete(Iterable<? extends T> entities) method on the ArrayList.

```
ArrayList arrayList = new ArrayList();
arrayList.add("catalog:engineering_as_a_service");
arrayList.add("catalog:engineeringasaservice");
repository.delete(arrayList);
```

Next, invoke the deleteAll() method, which makes use of the "all" view, on the repository instance.

```
repository.deleteAll();
```

All the entity instances get deleted as indicated by the Item Count of 0 in the Console in Figure 3-29.

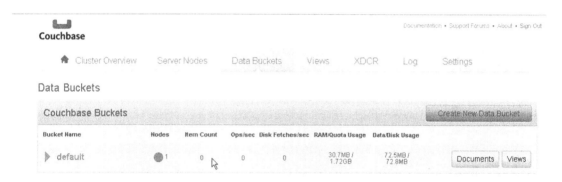

***Figure 3-29.*** *Item Count is 0 after deleting all documents*

Selecting Documents for the "default" bucket does not list any documents.
The CatalogService class is listed below.

```
package com.couchbase.service;

import java.io.IOException;
import java.net.URI;
import java.util.ArrayList;
import java.util.Iterator;
import java.util.LinkedList;
import java.util.List;
import org.springframework.beans.factory.annotation.Autowired;
import org.springframework.context.ApplicationContext;
import org.springframework.context.annotation.AnnotationConfigApplicationContext;
import com.couchbase.client.CouchbaseClient;
import com.couchbase.client.internal.HttpFuture;
import com.couchbase.client.protocol.views.DesignDocument;
import com.couchbase.client.protocol.views.ViewDesign;
```

```java
import com.couchbase.config.SpringCouchbaseApplicationConfig;
import com.couchbase.model.Catalog;
import com.couchbase.repositories.CatalogRepository;

public class CatalogService {
    // @Autowired
    public static CatalogRepository repository;

    public static void main(String[] args) {
        ApplicationContext context = new AnnotationConfigApplicationContext(
                SpringCouchbaseApplicationConfig.class);
        repository = context.getBean(CatalogRepository.class);

        List<URI> uris = new LinkedList<URI>();
        uris.add(URI.create("http://127.0.0.1:8091/pools"));
        CouchbaseClient couchbaseClient;
        try {
            couchbaseClient = new CouchbaseClient(uris, "default", "");
            DesignDocument designDoc = new DesignDocument("catalog");
            String viewName = "all";
            String mapFunction = "  function(doc,meta) {"
                    + "  if (meta.type == 'json') {"
                    + "  emit(doc.name, [doc.journal,doc.publisher,doc.edition,
                    doc.title,doc.author]);"
                    + "  }" + "}";
            String reduceFunction = "function(key, values, rereduce) {"
                    + "if (rereduce) {" + "var result = 0;"
                    + "for (var i = 0; i < values.length; i++) {"
                    + "result += values[i];" + "}" + "return result;"
                    + "  } else {" + " return values.length;" + "}" + "}";
            ViewDesign viewDesign = new ViewDesign(viewName, mapFunction,
                    reduceFunction);
            designDoc.getViews().add(viewDesign);
            // HttpFuture<java.lang.Boolean> httpFuture =
            // couchbaseClient.asyncCreateDesignDoc(designDoc);

        } catch (IOException e) {
            e.printStackTrace();
        }

        // countDocuments();
        // findAllDocuments();
        // findOneDocument();
        // findDocumentExists();
        // saveDocument();
        // saveDocuments();
        // deleteDocument();
        //deleteDocuments();
    }
```

```java
public static void countDocuments() {
    long count = repository.count();
    System.out.println("Number of catalogs: " + count);
}

public static void findAllDocuments() {
    Iterable<Catalog> iterable = repository.findAll();
    Iterator<Catalog> iter = iterable.iterator();
    while (iter.hasNext()) {
        Catalog catalog = iter.next();
        System.out.println("Journal: " + catalog.getJournal());
        System.out.println("Publisher: " + catalog.getPublisher());
        System.out.println("Edition: " + catalog.getEdition());
        System.out.println("Title: " + catalog.getTitle());
        System.out.println("Author: " + catalog.getAuthor());
    }

}

public static void findOneDocument() {
    Catalog catalog = repository
            .findOne("catalog:engineering-as-a-service");
    System.out.println("Journal: " + catalog.getJournal());
    System.out.println("Publisher: " + catalog.getPublisher());
    System.out.println("Edition: " + catalog.getEdition());
    System.out.println("Title: " + catalog.getTitle());
    System.out.println("Author: " + catalog.getAuthor());

}

public static void findDocumentExists() {

    boolean bool = repository
            .exists("catalog:quintessential-and-collaborative");
    System.out
            .println("Catalog with Id catalog:quintessential-and-collaborative exists: "
                    + bool);

}

public static void saveDocument() {

    Catalog catalog = new Catalog("Oracle Magazine", "Oracle Publishing",
            "11/12 2013", "Engineering as a Service", "Kelly, David");
    repository.save(catalog);

}
```

```
public static void saveDocuments() {

    ArrayList arrayList = new ArrayList();
    Catalog catalog1 = new Catalog("Oracle_Magazine", "Oracle_Publishing",
            "November-December_2013", "EngineeringasaService",
            "David_A._Kelly");
    Catalog catalog2 = new Catalog("Oracle_Magazine", "Oracle_Publishing",
            "11/12_2013", "Engineering_as_a_Service", "Kelly, David");
    arrayList.add(catalog1);
    arrayList.add(catalog2);
    repository.save(arrayList);

}

public static void deleteDocument() {
    // repository.delete("catalog:engineeringasaservice");

    Catalog catalog2 = new Catalog("Oracle_Magazine", "Oracle_Publishing",
            "11/12_2013", "Engineering_as_a_Service", "Kelly, David");
    repository.delete(catalog2);

}

public static void deleteDocuments() {
    ArrayList arrayList = new ArrayList();
    arrayList.add("catalog:engineering_as_a_service");
    arrayList.add("catalog:engineeringasaservice");
    // repository.delete(arrayList);
    repository.deleteAll();

}

}
```

# Summary

In this chapter we discussed using the Spring Data Couchbase project. We used Spring Data to access Couchbase Server with a Maven application and perform CRUD operations on the Couchbase Server. In the next chapter, we shall use PHP with Couchbase Server.

# CHAPTER 4

■ ■ ■

# Accessing Couchbase with PHP

PHP is one of the most commonly used scripting languages, and its usage (http://php.net/usage.php) for developing web sites continues to increase. PHP is an open source, object-oriented, server-side language and has the advantages of simplicity with support for all or most operating systems and web servers. The Couchbase PHP SDK provides access to Couchbase Server from a PHP script. JSON documents in Couchbase may be created and updated from PHP. In this chapter we shall use CRUD (create, read, update, delete) operations on Couchbase from PHP scripts. The chapter covers the following topics.

- Setting the Environment
- Installing PHP
- Installing the Couchbase PHP SDK
- Connecting with Couchbase Server
- Creating a Document
- Upserting a Document
- Getting a Document
- Replacing a Document
- Incrementing and Decrementing a Document
- Deleting a Document

## Setting the Environment

This chapter is based on 64-bit Windows OS. We need to install the following software in addition to installing Couchbase Server.

- PHP
- Web Server (packaged with PHP installation)
- Couchbase PHP SDK including C SDK (libcouchbase)

As the PHP and Couchbase PHP SDK binaries for Windows are specific to a Visual Studio version (Visual Studio 2012), download and install Visual C++ Redistributable for Visual Studio 2012, if not already installed, from https://www.microsoft.com/en-ca/download/details.aspx?id=30679. In the subsequent few sections we shall install the required software.

# Installing PHP

PHP 5.4 and later versions include a web server packaged in the PHP installation and does not require the web server to be installed separately. Download PHP 5.5 (5.5.26) VC11 x64 Thread-Safe version of the PHP zip file php-5.5.26-Win32-VC11-x64.zip from http://windows.php.net/download/. Extract the php-5.5.26-Win32-VC11-x64.zip file to a directory. A php-5.5.26-Win32-VC11-x64 directory gets created. Create a document root directory (C:\PHP used in this chapter) and copy the files and directories within the php-5.5.26-Win32-VC11-x64 directory to the C:/PHP directory. Rename the php.ini-development or php.ini-production in the root directory of the PHP installation, C:\PHP, to php.ini. Connect to the packaged web server at port 8000 with the following command from the document root directory C:\PHP directory.

```
php -S localhost:8000
```

The output from the command indicates that the Development Server has been started and listening on http://localhost:8000 as shown in Figure 4-1. PHP scripts copied to the document root directory can be run in the web server.

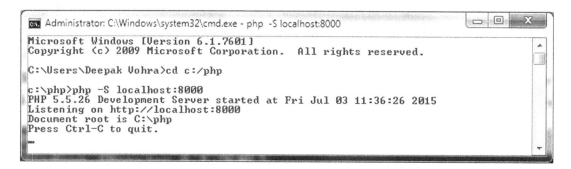

*Figure 4-1.* *Running the PHP Development Server*

To test that PHP and the web server have been installed, create the following PHP script index.php in the document root directory, the C:/PHP directory.

```php
<?php
 echo "<p>Welcome to PHP</p>";
?>
```

Run the PHP script on the web server by visiting the URL http://localhost:8000. The output from the script is shown in the browser in Figure 4-2.

*Figure 4-2.  Running the example Script index.php*

# Installing Couchbase PHP SDK

Next, we install the Couchbase PHP SDK.

1. Download PHP SDK prebuilt Windows binaries PHP 5.5 VC11 ZTS php_couchbase-2.0.7-5.5-zts-vc11-x64.zip file from http://docs.couchbase.com/developer/php-2.0/download-links.html. The thread-safe version (ZTS) should be downloaded and not the non-thread-safe version (NTS).

2. Extract the couchbase-2.0.7-5.5-zts-vc11-x64.zip file to a directory (for example, the C:/PHP directory).

3. Copy the libcouchbase.dll and the php_couchbase.dll from the C:\PHP\php_couchbase-2.0.7-5.5-zts-vc11-x64 directory to the C:\PHP directory.

4. Add the following line in the php.ini configuration file in the C:/PHP directory. Even though the libcouchbase.dll extension is required to be copied to the C:/PHP, it must not be added as an extension to the php.ini file as it is invoked internally by the php_couchbase.dll extension.

   **extension=php_couchbase.dll**

5. Shut down the PHP web server if running and restart the web server.

Next, we shall test if the Couchbase PHP SDK has been installed. Create a PHP script couchbaseconnect.php in the document root directory C:\PHP. Copy the following code listing to the couchbaseconnect.php script.

```php
<?php
 $cluster = new CouchbaseCluster('couchbase://localhost:8091');
 $bucket = $cluster->openBucket('default');
$res = $bucket->get('catalog');
var_dump($res);
 ?>
```

We shall be getting a document (added to Couchbase Server in an earlier chapter) using the Couchbase PHP SDK API. If a document with Id 'catalog' has not been retained from an earlier chapter log in to the Couchbase Admin Console and create a document with Id 'catalog' as shown in Figure 4-3 or use a document with some other Id (modify the preceding script to specify the document Id).

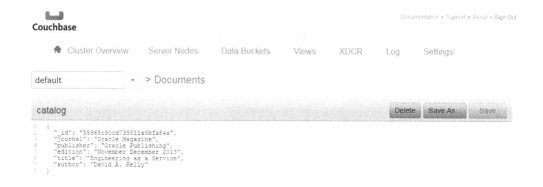

*Figure 4-3.* *Couchbase Server Document*

We shall discuss the script in a later section, but for testing run the script with the URL `http://localhost:8000/couchbaseconnect.php`. The script should generate the following output shown in Figure 4-4 in the browser to indicate that the Couchbase PHP SDK has been installed.

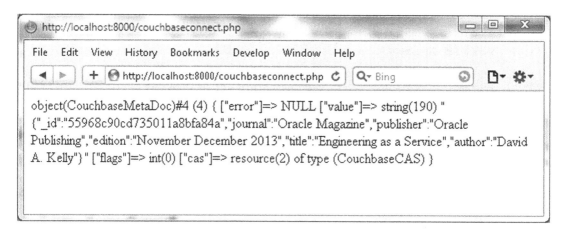

*Figure 4-4.* *Testing the Couchbase PHP SDK*

# Connecting with Couchbase Server

In this section we shall connect to Couchbase Server for which create a PHP script `connection.php` in the document root directory `C:\PHP`.

A connection to Couchbase Server cluster is represented with the `CouchbaseCluster` class in the Couchbase package. To create an instance of `CouchbaseCluster` class instance the class constructor with the following signature is provided.

```
__construct(string $dsn = 'http://127.0.0.1/', string $username = '', string $password = '')
```

The different parameters in the constructor are discussed in Table 4-1.

**Table 4-1.** *Constructors for the CouchbaseCluster Class*

| Parameter | Type | Description |
|---|---|---|
| $dsn | string | An string representing the Couchbase servers in the Couchbase cluster in the format `'couchbase://host1:port1,host2:port2,host3:port3'`. Specifying ports is optional if using standard ports. The ports on which server/s are running may be found from the Couchbase Administration Console. If one server is running the port is 8091. The hostname is localhost and the Ipv4 address of the server may also be specified, which is required to be specified if the server is running on a different machine than the PHP Client machine. If more than one server is specified in the string the first one that connects is used. |
| $username | string | The username to connect to Couchbase Server cluster. |
| $password | string | The password to connect to Couchbase Server cluster. |

Using the default hostname, port, and bucket, a Couchbase cluster connection can be obtained without specifying any connection parameters.

```
$cluster = new CouchbaseCluster();
```

Alternatively, the connection parameters may be specified redundantly, or if the connection parameters are other than the default the connection parameters must be specified. The hostname is localhost by default and port is 8091 by default. In the following CouchbaseCluster connection instance, the username and password are empty strings as username and password are not required for the default cluster.

```
$cluster = new CouchbaseCluster('couchbase://localhost:8091','','');
```

While the CouchbaseCluster class constructor creates a connection with a cluster to create a connection with a bucket the openBucket(string $name = 'default', string $password = '') method, which returns an instance of the CouchbaseBucket class, must be invoked to create a connection with a bucket. The CouchbaseBucket class provides various methods for adding, getting, updating and deleting data from a Couchbase bucket and also provides various properties to represent the connection state. Some of the methods in CouchbaseBucket class are discussed in Table 4-2.

**Table 4-2.** *Methods in the CouchbaseBucket Class*

| Method | Return Type | Description |
|---|---|---|
| manager() | CouchbaseBucketManager | Returns Couchbase bucket manager for management of the Couchbase bucket. |
| enableN1ql(mixed $hosts) | | Enables N1QL support. A cbq-server URI must be provided as an argument. |
| insert(string\|array $ids, mixed $val = NULL, array $options = array()) | mixed | Inserts a document. The method fails if the document already exists. |
| upsert(string\|array $ids, mixed $val = NULL, array $options = array()) | mixed | Like the insert method, inserts a document. But, unlike the insert method, updates the document if the document already exists. |
| replace(string\|array $ids, mixed $val = NULL, array $options = array()) | mixed | Replaces a document. |
| append(string\|array $ids, mixed $val = NULL, array $options = array()) | mixed | Appends data to a document. |
| prepend(string\|array $ids, mixed $val = NULL, array $options = array()) | mixed | Prepends data to a document. |
| remove(string\|array $ids, array $options = array()) | mixed | Removes a document. |
| get(string\|array $ids, array $options = array()) | mixed | Gets a document. |
| counter(string\|array $ids, integer $delta, array $options = array()) | mixed | Increments or decrements a document key by the specified $delta. |
| query(\CouchbaseQuery $query) | mixed | Queries a document based on a view query or N1QL query. |
| setTranscoder(string $encoder, string $decoder) | | Sets a custom encoder/decoder for serialization. |

Some of the properties in the CouchbaseBucket class are discussed in Table 4-3.

**Table 4-3.** *Properties in the CouchbaseBucket Class*

| Property | Type | Description |
|---|---|---|
| operationTimeout | integer | Operation timeout in milliseconds (ms) after which an operation is timed out. |
| viewTimeout | integer | View timeout in ms after which a view request is timed out. |
| httpTimeout | integer | HTTP timeout in ms. |
| configTimeout | integer | Configuration timeout after which the configuration is refreshed. |
| configNodeTimeout | integer | Configuration timeout for a node after which the configuration is refreshed. |

Obtain and output some of the bucket properties in the `connection.php` script. The `connection.php` script is listed below.

```php
<?php
    $cluster = new CouchbaseCluster();
$bucket = $cluster->openBucket('default');

echo "Operation timeout (ms): ";   echo $bucket->operationTimeout;
  echo "<br/>\n";
 echo "Configuration timeout (ms): ";   echo $bucket->configTimeout;
  echo "<br/>\n";
echo "Configuration Node timeout (ms) : ";   echo $bucket->configNodeTimeout;
  echo "<br/>\n";
echo "HTTP timeout (ms): ";   echo $bucket->httpTimeout;
  echo "<br/>\n";
echo "View timeout (ms): ";   echo $bucket->viewTimeout;
  echo "<br/>\n";
?>
```

Run the `connection.php` script with the URL `http://localhost:8000/connection.php`. The output from the script is shown in the browser in Figure 4-5.

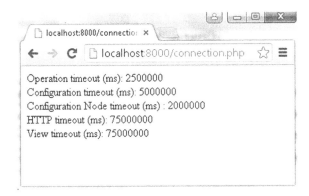

*Figure 4-5.* *Creating a Connection to Couchbase Server*

# Creating a Document

The `CouchbaseBucket` class provides two different methods to create a document in the Couchbase Server; `insert(string|array $ids, mixed $val = NULL, array $options = array())` and `upsert(string|array $ids, mixed $val = NULL, array $options = array())`. We shall discuss each of these methods in this section. The difference between the two methods is discussed in Table 4-2.

The difference between the two methods is that the `insert` method fails if the added document already exists and the `upsert` method updates the document if the document already exists.

Create a PHP script `insertDocument.php` in the `C:/PHP` directory. Create a Couchbase bucket connection instance to the 'default' bucket as discussed earlier.

```php
$cluster = new CouchbaseCluster();
$bucket = $cluster->openBucket('default');
```

105

The insert method is defined to accept different parameters, which are discussed in Table 4-4.

*Table 4-4.* *Parameters for the insert Method*

| Parameter | Type | Description |
|-----------|------|-------------|
| $ids | string\|array | The document IDs to store. |
| $val | mixed | The document value to store. |
| $options | array | An array of options. Supported options are expiry and flags. The expiry option is the time duration after which the document gets removed. By default the document does not get removed. |

At the minimum the document id and the document value are required to be specified. Use the insert method to create documents with ID catalog2. PHP includes the JSON extension, which provides the json_encode and json_decode methods to encode/decode JSON value respectively. The JSON extension is available in the PHP installation PHP 5.5 and is not required to be installed. A document with a JSON value may be stored by using an array to specify the key/value pairs for the JSON document.

```php
$arr = array('journal' => 'Oracle Magazine', 'publisher' => 'Oracle Publishing', 'edition'
=> 'November December  2013', 'title' => 'Quintessential and Collaborative', 'author' =>
'Tom Haunert');
```

Encode the array into a JSON representation using the json_encode method.

```php
$catalog2=json_encode($arr);
```

Store the JSON encoded array using the insert method.
```php
$res = $bucket->insert('catalog2', $catalog2);
```

The preceding insert method invocation creates document ID with JSON value. The insert method returns an object of type CouchbaseMetaDoc. To output the CAS value of the objects stored with the set method echo has been prepended to the set method statements. The insertDocument.php script is listed below.

```php
<?php
    $cluster = new CouchbaseCluster();
$bucket = $cluster->openBucket('default');

$arr = array('journal' => 'Oracle Magazine', 'publisher' => 'Oracle Publishing',
'edition' => 'November December  2013', 'title' =>

'Quintessential and Collaborative', 'author' => 'Tom Haunert');
$catalog2=json_encode($arr);
$res = $bucket->insert('catalog2', $catalog2);
var_dump($res);
?>
```

Run the PHP script in the browser with URL http://localhost:8000/insertDocument.php as shown in Figure 4-6.

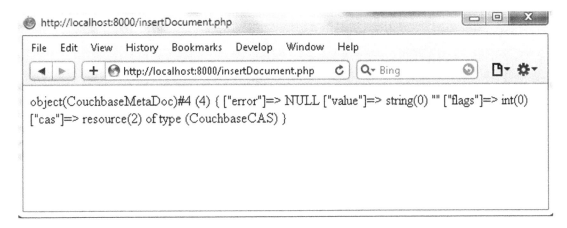

**Figure 4-6.** *Running the insertDocument.php Script*

In the Couchbase Server the document with the specified ID gets created as shown in the Couchbase Console in Figure 4-7.

**Figure 4-7.** *Listing Documents added with the insert Method*

Click on Edit Document for catalog2 document ID to display the JSON document for the ID as shown in Figure 4-8.

**Figure 4-8.** *Displaying JSON for a Document catalog2*

If the `insertDocument.php` script is run again without deleting the `catalog2` document, the *Uncaught exception 'CouchbaseException' with message 'The key already exists in the server.'* exception gets generated.

# Upserting a Document

The `upsert(string|array $ids, mixed $val = NULL, array $options = array())` method is used to add a new document or update an existing one. The `upsert` method is defined to accept the same parameters as the `insert` method.

Create a PHP script `upsertDocument.php` in the document root directory C:/PHP. Create JSON encoded array of catalog fields. Use the same document id 'catalog2' and the same keys as the `catalog2` document added earlier but supply different values. Invoke the `upsert` method on the JSON encoded array using a CouchbaseBucket instance to store the updated document.

```
$arr = array('journal' => 'OracleMagazine', 'publisher' => 'OraclePublishing', 'edition' =>
'11/12  2013', 'title' => 'Quintessential and Collaborative', 'author' => 'Haunert, Tom');
$catalog2=json_encode($arr);$res = $bucket->upsert('catalog2', $catalog2);
```

The PHP script `upsertDocument.php` is listed below.

```
<?php
    $cluster = new CouchbaseCluster();
$bucket = $cluster->openBucket('default');$arr = array('journal' => 'OracleMagazine',
'publisher' => 'OraclePublishing', 'edition' => '11/12  2013', 'title' => 'Quintessential
and Collaborative', 'author' => 'Haunert, Tom');
$catalog2=json_encode($arr);
$res = $bucket->upsert('catalog2', $catalog2);
 var_dump($res);
?>
```

Run the `upsertDocument.php` script with the URL `http://localhost:8000/upsertDocument.php` as shown in Figure 4-9.

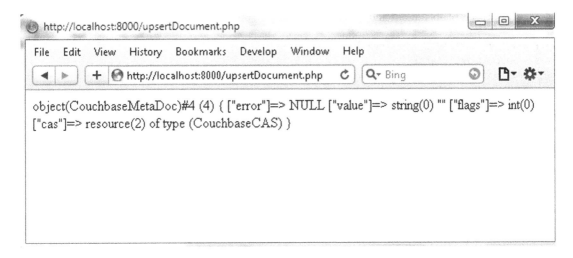

*Figure 4-9. Running the upsertDocument.php Script*

The JSON document for document ID `catalog2` added gets updated as shown in Figure 4-10.

*Figure 4-10. Listing Updated Document*

# Getting a Document

In this section we shall retrieve document/s previously added to the server. Create a PHP script getDocument.php in the C:/PHP folder. The get(string|array $ids, array $options = array()) method is used to get a document.

The get method returns the document object. The get method is defined to accept different parameters, which are discussed in Table 4-5.

*Table 4-5. Parameters for the get Method*

| Parameter | Type | Description |
| --- | --- | --- |
| $ids | string\|array | The document id/s to get. |
| $options | array | An array of options. |

Create an array of documents and invoke the get method on the array.

```php
$arr = array('catalog', 'catalog2');
$res = $bucket->get($arr);
```

The PHP script getDocument.php is listed below.

```php
<?php
    $cluster = new CouchbaseCluster();
$bucket = $cluster->openBucket('default');

$arr = array('catalog', 'catalog2');
$res = $bucket->get($arr);
var_dump($res);
?>
```

Run the PHP script with URL http://localhost:8000/getDocument.php. The documents get retrieved and output in the browser including the CAS value as shown in Figure 4-11.

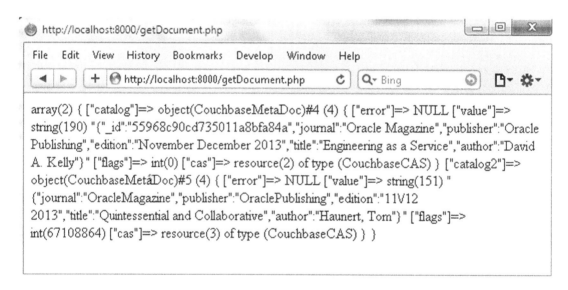

***Figure 4-11.*** *Running the getDocument.php Script*

A single document may be output by providing the document id. For example, the document with id 'catalog' is output as follows.

```php
var_dump($res["catalog"]);
```

When the getDocument.php script is run again, just the 'catalog' id document is output as shown in Figure 4-12.

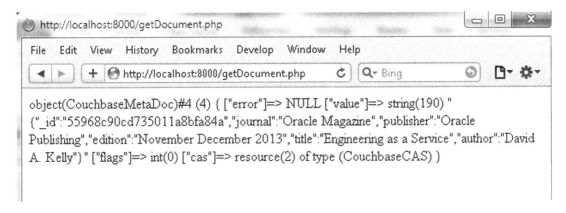

object(CouchbaseMetaDoc)#4 (4) { ["error"]=> NULL ["value"]=> string(190) " {"_id":"55968c90cd735011a8bfa84a","journal":"Oracle Magazine","publisher":"Oracle Publishing","edition":"November December 2013","title":"Engineering as a Service","author":"David A. Kelly"}" ["flags"]=> int(0) ["cas"]=> resource(2) of type (CouchbaseCAS) }

*Figure 4-12. Running the getDocument.php Script to access a single document*

# Replacing a Document

The CouchbaseBucket class provides the replace(string|array $ids, mixed $val = NULL, array $options = array()) method for replacing a document. The method parameters are the same as those of the insert and the upsert methods. The $options parameter supports an additional option of CAS.

Create a script replaceDocument.php in the C:/PHP folder. Specify the JSON encoded replacement document.

```
$arr = array('journal' => 'Oracle Magazine', 'publisher' => 'Oracle Publishing', 'edition'
=> 'November December  2013', 'title' => 'Quintessential and Collaborative', 'author' =>
'Tom Haunert');
$catalog2=json_encode($arr);
```

Invoke the replace method to replace the document ID 'catalog.'

```
$res = $bucket->insert('catalog2', $catalog2);
```

The replaceDocument.php script is listed below.

```
<?php
    $cluster = new CouchbaseCluster();
$bucket = $cluster->openBucket('default');
$arr = array('journal' => 'Oracle Magazine', 'publisher' => 'Oracle Publishing', 'edition'
=> 'November December  2013', 'title' => 'Quintessential and Collaborative', 'author' =>
'Tom Haunert');
$catalog2=json_encode($arr);
$res = $bucket->insert('catalog2', $catalog2);
 var_dump($res);
?>
```

Run the PHP script with the URL http://localhost:8000/replaceDocument.php as shown in Figure 4-13.

111

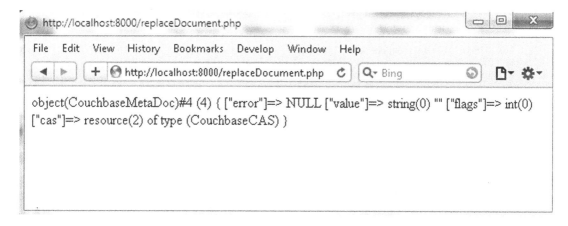

*Figure 4-13. Running the replaceDocument.php Script*

In the Couchbase Console the document is shown to be replaced for the specified document ID as shown in Figure 4-14. The catalog2 document, which we had upserted using the upsert method, gets replaced with the document before the upsert.

*Figure 4-14. Listing Replaced Document*

## Incrementing and Decrementing a Document

The CouchbaseBucket class provides the counter(string|array $ids, integer $delta, array $options = array()) method to increment/decrement the numeric value of a document in the cluster.

The counter method is defined to accept different parameters, which are discussed in Table 4-6.

***Table 4-6.*** *Parameters for the counter Method*

| Parameter | Type | Description |
|---|---|---|
| $ids | string\|array | The document IDs whose value is to be incremented/decremented. The value must be numeric. |
| $delta | integer | The value to increment/decrement by. Default is 1. If a negative value is supplied, the value is decremented. |
| $options | array | Array of options. Supported options are initial and expiry. The initial option may be used to specify an initial value for the document and to create the document if the document does not already exist. |

Create a PHP script counterDocument.php in the C:/PHP folder. To demonstrate the increment method, add a key-value pair with numeric value 15 using the insert method. Subsequently increment the value by 5.

```
$bucket->insert('id1', 15);
$res=$bucket->counter('id1',5);
```

To demonstrate argument decrementing a document value, create a key/value pair with value as 1 and subsequently invoke the counter method with $delta as -1.

```
$bucket->insert('id2', 1);
$res= $bucket->counter('id2',-1);
```

To demonstrate setting the initial value for a document, specify the initial option in the counter method. Set the initial option to a value of 10 .

```
$res=$bucket->counter('id3', 5, array('initial'=>10));
```

If the initial option argument is used to create a document the document value does not get incremented/decremented in the same invocation of the counter method. The counter method must be invoked again to increment/decrement the numeric value.

```
$res=$bucket->counter('id3', 5);
```

Couchbase stores numbers as unsigned values. Negative numbers cannot be incremented using the counter method if after the increment the value is still negative. An integer overflow occurs and a non-logical numerical result is returned. For example, set a document to –15. Subsequently increment the document by 5.

```
$bucket->insert('id4', -15);
$res=$bucket->counter('id4',5);
```

When the script is run the value is not incremented to –10 as expected but a non-logical numerical value is stored due to an integer overflow. A value cannot be decremented below 0. For example store document 'id5' with a value of 1 and subsequently decrement using a delta of -2. Instead of -1 the value is decremented to 0.

```
$bucket->insert('id5', 1);
$res=$bucket->counter('id5',-2);
```

The counter.php script is listed below.

```php
<?php
    $cluster = new CouchbaseCluster();
$bucket = $cluster->openBucket('default');

$bucket->insert('id1', 15);
$res=$bucket->counter('id1',5);
var_dump($res);

$bucket->insert('id2', 1);
$res= $bucket->counter('id2',-1);
var_dump($res);
$res=$bucket->counter('id3', 5, array('initial'=>10));
var_dump($res);
$res=$bucket->counter('id3', 5);
var_dump($res);
$bucket->insert('id4', -15);
$res=$bucket->counter('id4',5);
var_dump($res);
$bucket->insert('id5', 1);
$res=$bucket->counter('id5',-2);
var_dump($res)?>
```

Run the counter.php in a browser with URL http://localhost:8000/counter.php as shown in Figure 4-15.

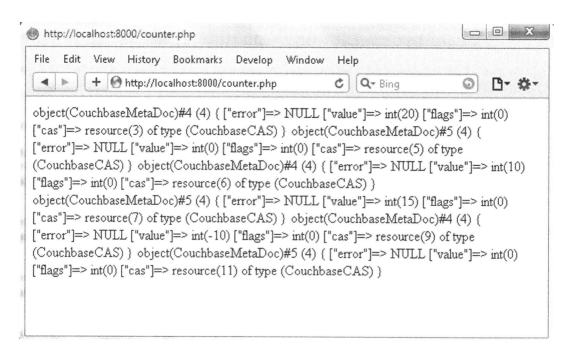

*Figure 4-15. Running the counter.php Script*

The integer 15 incremented by 5 is stored as 20 as shown in Figure 4-16. The document 'id3', which was initialized to 10 and incremented in the same invocation of `counter` by 5 does not get incremented until the `counter` method is invoked again to increment by 5. The stored value is 15 and not 20, which would have been the stored value if the `counter` method had also incremented in the first invocation that included the initialization to 10. Document `id4`, which is initially -15 and incremented by 5 is stored as a non-logical number due to an integer overflow. Document `id5`, which has an initial value of 1, does not get decremented below 0 even though a delta of -2 is applied.

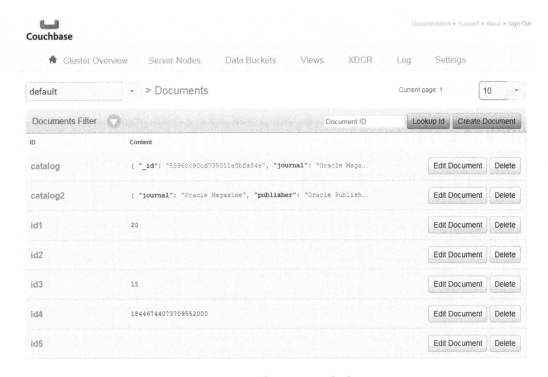

***Figure 4-16.*** *Result of incrementing/decrementing with counter method*

The 'id2' document that has the value of 1 to start with gets decremented to 0 after a delta of -1 is applied as shown in Figure 4-17. Similarly the 'id5' document is decremented to 0 to demonstrate that a document value does not get decremented below 0 even if the delta applied makes the initial value a negative integer.

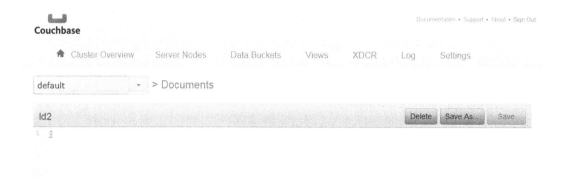

***Figure 4-17.*** *Result of decrementing with Negative Delta*

# Deleting a Document

In this section we shall delete a document using the remove(string|array $ids, array $options = array()) method in the CouchbaseBucket class.

The remove method provides the parameters listed in Table 4-7.

***Table 4-7.*** *Parameters for the remove Method*

| Method | Type | Description |
| --- | --- | --- |
| $ids | String|array | The document ids to delete. |
| $options | array | An array of options. The only supported option is CAS. |

Create a PHP script removeDocument.php in the C:/PHP folder. Create an array of documents to remove and invoke the remove method. The removeDocument.php script is listed below.

```php
<?php
    $cluster = new CouchbaseCluster();
bucket = $cluster->openBucket('default');
$arr = array('catalog2','id1','id2','id3','id4','id5');
$res = $bucket->remove($arr);
 var_dump($res);
?>
```

Run the PHP script with URL http://localhost:8000/removeDocument.php as shown in Figure 4-18.

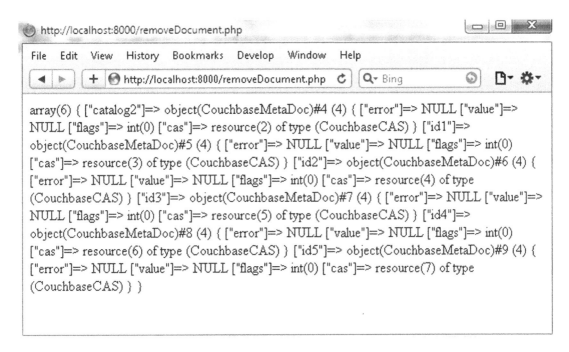

array(6) { ["catalog2"]=> object(CouchbaseMetaDoc)#4 (4) { ["error"]=> NULL ["value"]=>
NULL ["flags"]=> int(0) ["cas"]=> resource(2) of type (CouchbaseCAS) } ["id1"]=>
object(CouchbaseMetaDoc)#5 (4) { ["error"]=> NULL ["value"]=> NULL ["flags"]=> int(0)
["cas"]=> resource(3) of type (CouchbaseCAS) } ["id2"]=> object(CouchbaseMetaDoc)#6 (4) {
["error"]=> NULL ["value"]=> NULL ["flags"]=> int(0) ["cas"]=> resource(4) of type
(CouchbaseCAS) } ["id3"]=> object(CouchbaseMetaDoc)#7 (4) { ["error"]=> NULL ["value"]=>
NULL ["flags"]=> int(0) ["cas"]=> resource(5) of type (CouchbaseCAS) } ["id4"]=>
object(CouchbaseMetaDoc)#8 (4) { ["error"]=> NULL ["value"]=> NULL ["flags"]=> int(0)
["cas"]=> resource(6) of type (CouchbaseCAS) } ["id5"]=> object(CouchbaseMetaDoc)#9 (4) {
["error"]=> NULL ["value"]=> NULL ["flags"]=> int(0) ["cas"]=> resource(7) of type
(CouchbaseCAS) } }

***Figure 4-18.*** *Running the removeDocument.php Script*

In the Couchbase Console the documents with ids 'catalog2', 'id1', 'id2', 'id3', 'id4', and 'id5' are deleted, and as shown in Figure 4-19 the only document listed is the 'catalog' id document, which was not included in the array of documents to remove.

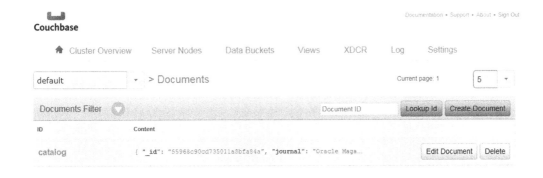

***Figure 4-19.*** *The deleted documents are not listed*

# Summary

In this chapter we discussed using Couchbase Server with PHP. In the next chapter we shall discuss using Couchbase Server with Ruby, including accessing the server and performing CRUD operations.

▓ ▓ ▓

# CHAPTER 5

# Accessing with Ruby

Ruby is an object-oriented, open source, programming language. Some of the salient features of Ruby are simplicity, flexibility, extensibility, portability, and OS independent threading. The Ruby Client library for Couchbase provides access to Couchbase server from a Ruby application. In this chapter we shall access Couchbase server using Ruby and perform simple CRUD (create, read, update, delete) operations. The chapter covers the following topics.

- Setting the Environment
- Installing Ruby
- Installing DevKit
- Installing Ruby Client for Couchabse
- Connecting with Couchbase Server
- Creating a Document in Couchbase Server
- Retrieving a Document
- Updating a Document
- Deleting a Document
- Querying a Document with a View

## Setting the Environment

We need to download and install the following software to access Couchbase Server from Ruby.

1. Ruby Installer for Ruby 2.1.6 (rubyinstaller-2.1.6-x64.exe). Download it from `http://rubyinstaller.org/downloads/`. The Couchbase Ruby client library supports Ruby versions 1.8.7, 1.9.3, 2.0, and 2.1.

2. Rubygems.

3. RubyInstaller Development Kit (DevKit). Download the appropriate version from `http://rubyinstaller.org/downloads/`. For Ruby 2.1.6, download DevKit-mingw64-64-4.7.2-20130224-1432-sfx.exe.

4. Ruby Client Library.

The specific versions listed for Ruby and DevKit are the versions to download. Ruby Client for Couchbase may not get installed with other versions.

# Installing Ruby

To install Ruby double-click on the Ruby installer application rubyinstaller-2.1.6-x64.exe. The Ruby Setup wizard starts.

1. Select the Setup Language and click on OK.

2. Accept the License Agreement and click on Next.

3. In Installation Destination and Optional Tasks specify a destination folder to install Ruby, or select the default folder. The directory path should not include any spaces. Select Add Ruby Executables to your PATH.

4. Click on Install as shown in Figure 5-1.

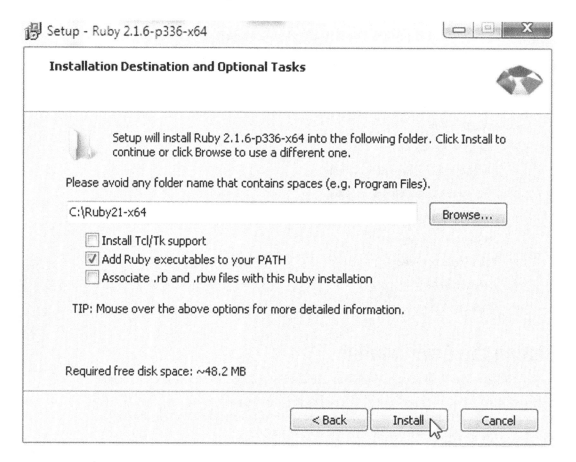

***Figure 5-1.*** *Selecting Installation Directory for Ruby*

Ruby starts installing as shown in Figure 5-2.

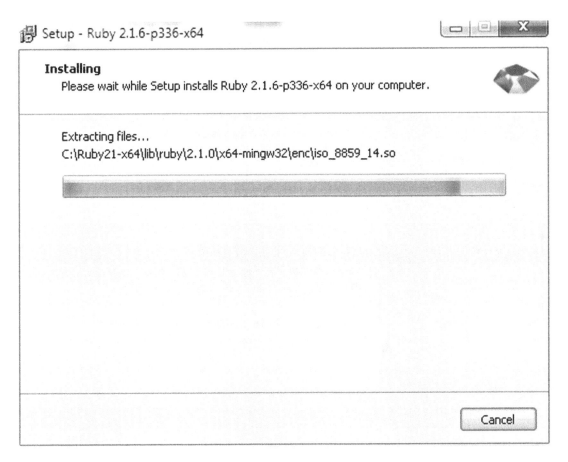

*Figure 5-2.* *Installing Ruby*

The Setup wizard completes installing Ruby as shown in Figure 5-3. Click on Finish.

**Figure 5-3.** *Ruby Installed*

Next, install Rubygems, which is a package management framework for Ruby. Run the following command to install Rubygems.

```
gem install rubygems-update
```

The rubygems gem gets installed as shown in Figure 5-4.

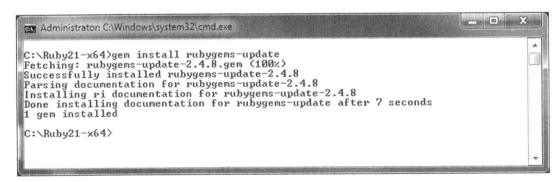

**Figure 5-4.** *Installing Rubygems*

If Rubygems is already installed update to the latest version with the following command.

```
update_rubygems
```

Rubygems gets updated as shown in Figure 5-5.

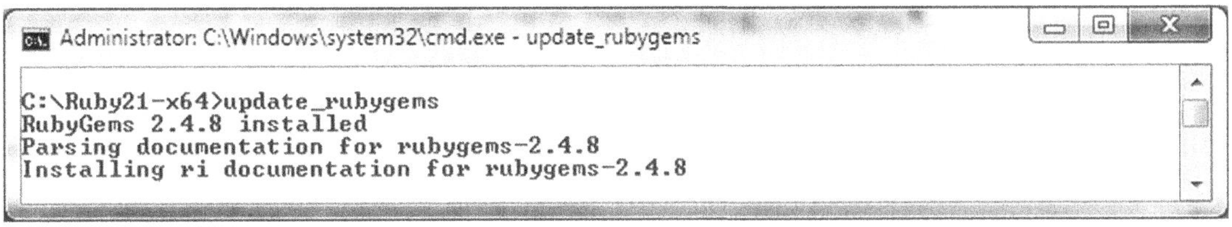

```
Administrator: C:\Windows\system32\cmd.exe - update_rubygems

C:\Ruby21-x64>update_rubygems
RubyGems 2.4.8 installed
Parsing documentation for rubygems-2.4.8
Installing ri documentation for rubygems-2.4.8
```

***Figure 5-5.*** *Updating Rubygems*

# Installing DevKit

DevKit is a toolkit that is used to build many of the C/C++ extensions available for Ruby. To install DevKit double-click on the DevKit installer application (DevKit-mingw64-64-4.7.2-20130224-1432-sfx.exe) to extract DevKit files to a directory. cd (change directory) to the directory in which the files are extracted.

```
cd C:\Ruby21-x64
```

Initialize DevKit and auto-generate the config.xml file using the following command.

```
ruby dk.rb init
```

A config.xml file gets generated in the C:\Ruby21-x64 directory. Add the following line to the config.xml file.

```
- C:/Ruby21-x64
```

Install DevKit using the following command.

```
ruby dk.rb install
```

Verify that DevKit has been installed using the following command.

```
ruby -rubygems -e "require 'json'; puts JSON.load('[42]').inspect"
```

The output from the preceding commands to initialize/install DevKit are shown in command shell in Figure 5-6.

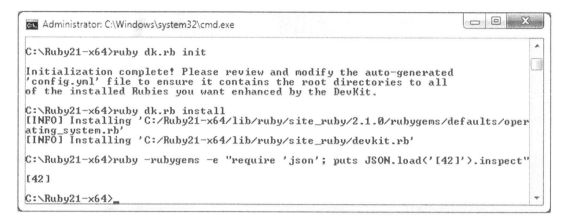

*Figure 5-6.* *Installing Devkit*

# Installing Ruby Client Library

Couchbase C Client library (libcouchbase) is a pre-requisite for installing the Ruby Client library, but for Windows the dependencies are included in the couchbase gem. Run the following command to install the Ruby Client library for Couchbase.

```
gem install couchbase
```

The Ruby Client library gem for couchbase and the dependencies get installed as shown in Figure 5-7.

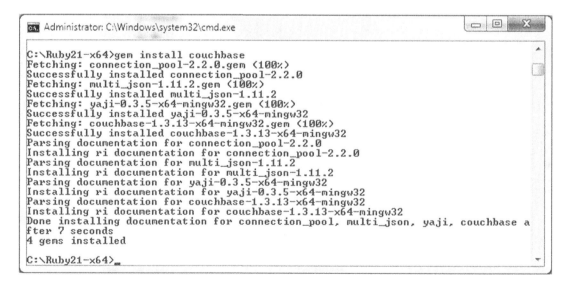

*Figure 5-7.* *Installing Ruby Client Library for Couchbase*

To test that the Ruby Client library has been installed create a Ruby file (couchbaseconnect.rb for example) and add the following script to the file.

```
require 'rubygems'
require 'couchbase'
client = Couchbase.connect("http://localhost:8091")
client.set("Client Type","Ruby")
```

Run the following command in a command shell to run the script.

```
ruby couchbaseconnect.rb
```

Connection with Couchbase Server gets established from the Ruby script and a key-value pair gets created in Couchbase Server as shown in the Couchbase Console in Figure 5-8. The Ruby script does not generate an output message.

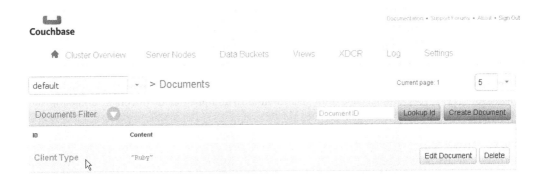

***Figure 5-8.***  *Creating a Key/Value Pair in Couchbase Server*

# Connecting with Couchbase Server

Create a Ruby script connection.rb to connect to Couchbase Server. First, we need to load the Ruby Client for Couchbase library and the rubygems using require statements. At the minimum a Couchbase.connect or Couchbase.new method invocation is required to establish a connection with Couchbase Server.

```
require 'rubygems'
require 'couchbase'
client = Couchbase.connect
client2= Couchbase.new
```

We did not need to specify any connection URL or port. The default endpoint used to connect to Couchbase Server is http://localhost:8091/pools/default/buckets/default. Run the script with the following command.

```
ruby connection.rb
```

The script runs and a connection with Couchbase Server gets established as shown in Figure 5-9.

```
Command Prompt                                                    _ □ ×
C:\Couchbase\Ruby>ruby connection.rb

C:\Couchbase\Ruby>_
```

***Figure 5-9.*** *Running the connection.rb Ruby Script*

The default endpoint connects to the default pool and the default bucket. The pool and the bucket may also be specified explicitly using the :pool and :bucket ruby connect options.

```
client = Couchbase.connect(:pool => "default", :bucket => "default")
```

The default bucket is not password protected and does not require a username or password to be specified, but if using another bucket the username and password may be specified.

```
client =Couchbase.connect(:bucket => 'catalog',
                :username => 'user',
                :password => 'password')
```

If the 'catalog' bucket is not defined in Couchbase server but specified in the connection the Couchbase::Error::BucketNotFound error gets created. To demonstrate do not create a 'catalog' bucket but specify the bucket in the connection.

```
require 'rubygems'
require 'couchbase'
client = Couchbase.connect(:bucket => "catalog")
#client = Couchbase.connect("http://localhost:8091","catalog")
```

The Couchbase::Error::BucketNotFound error gets generated as shown in Figure 5-10.

```
C:\Couchbase\Ruby>ruby connection.rb
C:/Ruby200-x64/lib/ruby/gems/2.0.0/gems/couchbase-1.3.4-x64-mingw32/lib/couchbas
e.rb:63:in `initialize':  (error=0x19) (Couchbase::Error::BucketNotFound)
        from C:/Ruby200-x64/lib/ruby/gems/2.0.0/gems/couchbase-1.3.4-x64-mingw32
/lib/couchbase.rb:63:in `new'
        from C:/Ruby200-x64/lib/ruby/gems/2.0.0/gems/couchbase-1.3.4-x64-mingw32
/lib/couchbase.rb:63:in `connect'
        from connection.rb:37:in `<main>'

C:\Couchbase\Ruby>
```

***Figure 5-10.*** *Couchbase::Error::BucketNotFound*

By default hostname is 'localhost' and may also be explicitly specified as 'localhost'. The hostname may also be specified explicitly as '127.0.0.1' or as the Ipv4 address using the :hostname option. By default the port is 8091 and the port may be explicitly specified using the :port option.

```
client = Couchbase.connect(:hostname => "127.0.0.1", :port => 8091)
```

All of the following are alternative methods of connecting to Couchbase Server on the default bucket and the default host and port; 127.0.0.1 is the Couchbase Server node name, 192.168.1.71 is the Ipv4 address for the localhost and 8091 is the default port number. The Ipv4 address would be different for different users/systems and should be replaced in the following script and other scripts in which it is used. The Ipv4 address may be found with the ipconfig /all command. The user credentials could be different than those listed.

```
client = Couchbase.connect("http://localhost:8091")
c3 = Couchbase.connect("http://127.0.0.1:8091/pools/default/buckets/default")
c4 = Couchbase.connect("http://127.0.0.1:8091/pools/default")
c5 = Couchbase.connect("http://127.0.0.1:8091")
c6 = Couchbase.connect(:hostname => "127.0.0.1")
c7 = Couchbase.connect(:hostname => "127.0.0.1", :port => 8091)
c8 = Couchbase.connect(:pool => "default", :bucket => "default")
c10 = Couchbase.connect("http://127.0.0.1:8091","default")
c12 =Couchbase.connect('http://192.168.1.71:8091/pools/default/buckets/default',
                  :username => 'default',
                  :password => '')
```

A list of possible nodes can be specified using the :node_list option.

```
c9 = Couchbase.connect(:bucket => "default",
                   :node_list => ['127.0.0.1:8092', '127.0.0.1:8091'])
```

If one of the nodes is not available another node is tried from the node list till a connection is established. After a connection is established the node list is not referred to again and the current cluster topology is used to re-connect on failover or rebalance. By default whenever a connection is requested a new connection instance is created, but the Couchbase.bucket method can be used to create a persistent connection instance and store the connection instance in memory. When a connection is requested again the connection instance stored in the thread storage is returned.

```
client=Couchbase.bucket
```

The following Ruby script connection.rb includes alternative methods of connecting to Couchbase Server. To test that a connection gets established with each of these alternatives add a key-value pair for each the alternative connection instances.

```
require 'rubygems'
require 'couchbase'
client = Couchbase.connect("http://localhost:8091")
c = Couchbase.connect
c.set("c","ruby client")
c2 = Couchbase.new
c2.set("c2","ruby client")
c3 = Couchbase.connect("http://127.0.0.1:8091/pools/default/buckets/default")
c3.set("c3","ruby client")
c4 = Couchbase.connect("http://127.0.0.1:8091/pools/default")
c4.set("c4","ruby client")
c5 = Couchbase.connect("http://127.0.0.1:8091")
c5-set("c5","ruby client")
c6 = Couchbase.connect(:hostname => "127.0.0.1")
c6.set("c6","ruby client")
c7 = Couchbase.connect(:hostname => "127.0.0.1", :port => 8091)
c7.set("c7","ruby client")
```

```
c8 = Couchbase.connect(:pool => "default", :bucket => "default")
c8.set("c8","ruby client")
c9 = Couchbase.connect(:bucket => "default",
                       :node_list => ['127.0.0.1:8092', '127.0.0.1:8091'])
c9.set("c9","ruby client")
c10 = Couchbase.connect("http://127.0.0.1:8091","default")
c10.set("c10","ruby client")
c11 =Couchbase.connect(:bucket => 'default',
                  :username => '',
                  :password => '')
c11.set("c11","ruby client")
c12 =Couchbase.connect('http://192.168.1.71:8091/pools/default/buckets/default',
                  :username => 'default',
                  :password => '')
c12.set("c12","ruby client")
c1 =Couchbase.bucket
c1.set("c1","ruby client")
```

Run the script with the following command.

```
ruby connection.rb
```

The key-value pairs for the different alternative methods of creating a connection instance are created in Couchbase Server as shown in Couchbase Console in Figure 5-11.

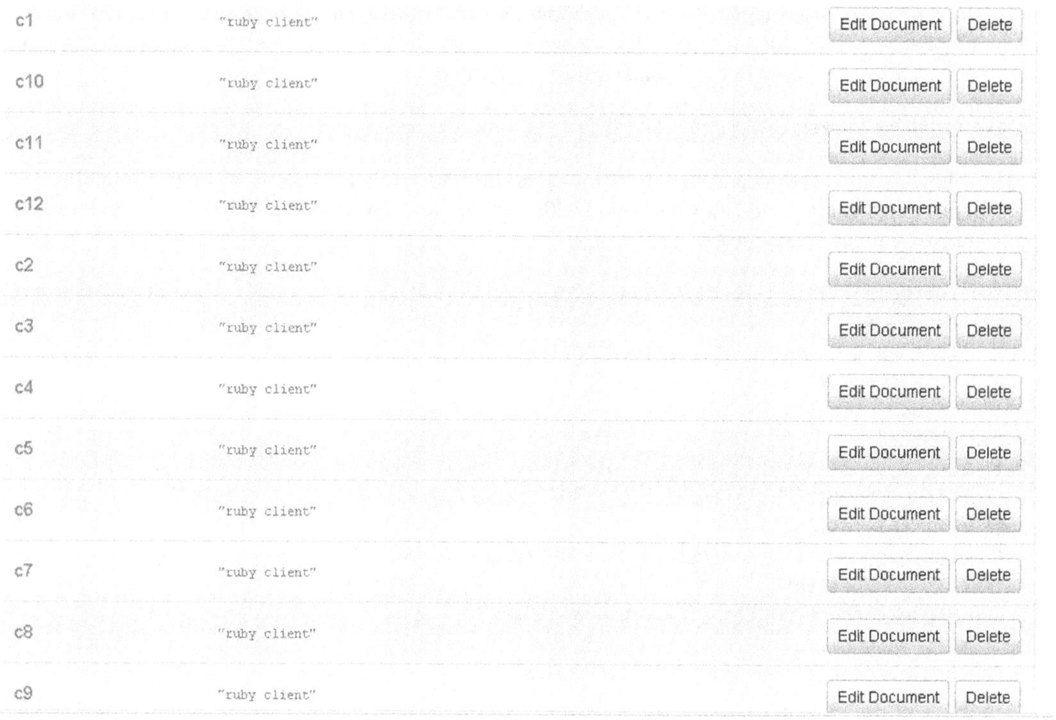

*Figure 5-11.* *Key/value Pairs in Couchbase Server*

A connection instance can be created as a Singleton object using `connection_options`.

```
Couchbase.connection_options = {:bucket => "default",:hostname => "127.0.0.1",
:password => ""}
```

The `connection_options` may even be empty initially and the connection options may be set using the set method.

```
client =Couchbase.bucket.set("c1","value set using Couchbase.bucket.set")
```

Or, the connection options of the singleton connection may even be modified at runtime, on reconnect.

```
Couchbase.bucket.reconnect(:bucket => "catalog")
```

In the following Ruby script the value of c1 key is updated using the set method.

```
require 'rubygems'
require 'couchbase'
Couchbase.connection_options = {:bucket => "default",:hostname => "127.0.0.1",
:password => ""}
 c1 =Couchbase.bucket.set("c1","value set using Couchbase.bucket.set")
```

When the script is run the value of key c1 gets updated as shown in the Couchbase Console as shown in Figure 5-12.

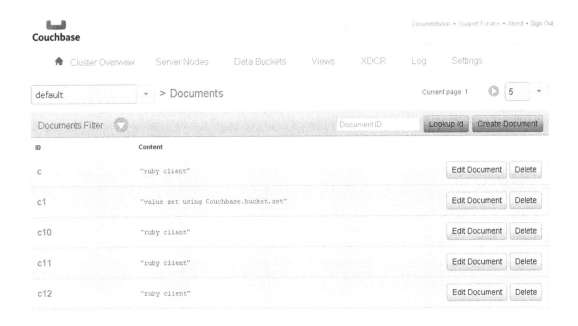

***Figure 5-12.*** *Updating value of key c1*

Timeout for a connection can be set using the :timeout option. The timeout can later be set to another value using the timeout property. In the following connection.rb script the timeout is initially set to 10 second and subsequently modified to 1 x 10^-6 seconds.

```
require 'rubygems'
require 'couchbase'
 conn = Couchbase.connect(:timeout => 10_000_000)
 conn.timeout = 1_0
 conn.set("connection timeout", "1_0")
```

When the script is run the Couchbase:Error:Timeout error is generated because the connection times out before the connection is established as shown in Figure 5-13.

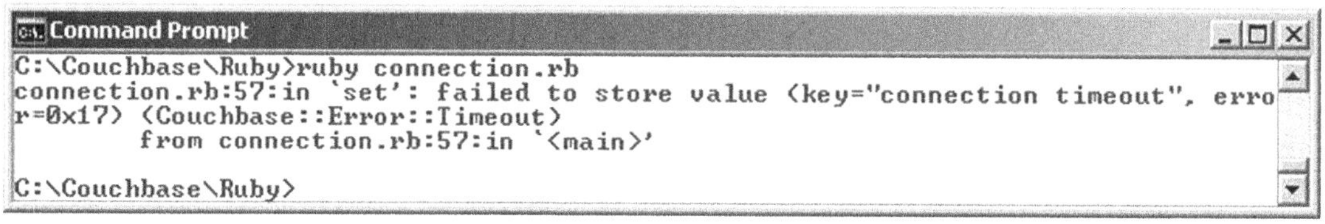

*Figure 5-13.* *Couchbase:Error:Timeout error*

# Creating a Document in Couchbase Server

To create and store a JSON object in Couchbase Server two methods are available, set and add. The difference between set and add is that the set method overrides the value if the same key is already defined in the server, while the add method can only be used to add a new key-value pair and if used to add a key already defined generates an error.

## Setting a Document

The set method is used to create a new key-value pair or set the value of a key already in the server to a new value. The set method has the following signature.

```
object.set(key, value, options)
```

The set method returns the CAS value of the object stored as a fixed number. The method args are discussed in Table 5-1.

*Table 5-1.* *The set method Arguments*

| Argument | Type | Description |
|---|---|---|
| key | string | Document ID used to identify the value. Must be unique in a bucket. |
| value | object | Value to be stored. |
| options | hash | Options for the set method. The :format option is used to specify the format of the stored value. The different formats supported are :document for JSON data, :plain for string storage and :marshal to serialize the ruby object using Marshal.dump and Marshal.load. The :cas option can be used to set the CAS value of the object stored, which is unique and generated by the server. The :cas may be used to perform optimistic concurrency control while setting a value and is discussed with an example in this section. |

Create a Ruby script storeDocument.rb and use the set method to set different key-value pairs. For example, set key-value pairs for journal, publisher, edition, title and author. Also set a key-value pair with the value as a JSON document.

```
require 'rubygems'
require 'couchbase'
client = Couchbase.connect("http://127.0.0.1:8091")
client.set("journal","Oracle Magazine")
client.set("publisher","Oracle Publishing")
client.set("edition","November-December 2013")
client.set("title","Quintessential and Collaborative")
client.set("author","Tom Haunert")
client.set("catalog2","{'journal': 'Oracle Magazine','publisher': 'Oracle
Publishing','edition': 'November December 2013','title': 'Engineering as a
Service','author': 'David A. Kelly'}")
```

Run the script with the following command.

```
ruby storeDocument.rb
```

The different key-value pairs get set as shown in the Couchbase Console as shown in Figure 5-14.

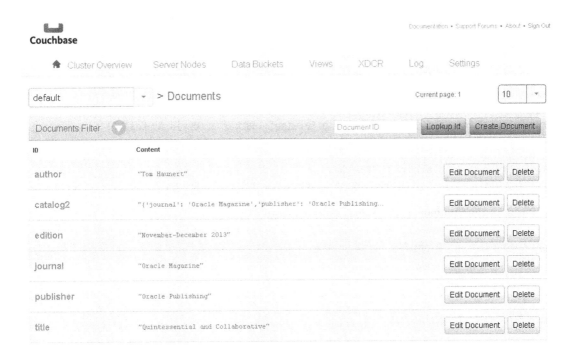

***Figure 5-14.*** *Key/Value Pairs added with the set method*

The `catalog2` key has a JSON object as a value as shown in Figure 5-15.

***Figure 5-15.*** *Key/Value Pair for key catalog2 has JSON document as value*

Next, we demonstrate optimistic concurrency control in setting a value. The set method returns the CAS value of the object stored on the server. For optimistic concurrency control the last known CAS value may be used when invoking the `set` method on the same key again. Update the `storeDocument.rb` script to store the CAS value returned by the `set` method in a variable and subsequently specify the CAS value in the `:cas` option when invoking the `set` method again.

```
require 'rubygems'
require 'couchbase'
client = Couchbase.connect("http://127.0.0.1:8091")
cas=   client.set("catalog2","{'journal': 'Oracle Magazine','publisher': 'Oracle
Publishing','edition': 'November
December 2013','title': 'Engineering as a Service','author': 'David A. Kelly'}")
print cas
client.set("catalog2", "{'journal': 'Oracle Magazine','publisher':
'Oracle-Publishing','edition': 'November
December 2013','title': 'Engineering as a Service','author': 'Kelly, A. David'}", :cas=>cas)
```

Run the script to output the CAS value as shown in Figure 5-16.

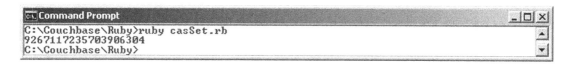

***Figure 5-16.*** *Outputting a Cas Value*

The value of the key `catalog2` is first set to a JSON object and the CAS value for the object returned and stored in a variable in the script. Subsequently the CAS value is used to re-set the value of the `catalog2` ID to a different value as shown in Figure 5-17. Concurrency control is used to avoid data corruption when multiple operations are run concurrently. With the `:cas` option it is ensured that no other operation modifies the value of a key while the `set` method using the CAS value runs.

*Figure 5-17.* *JSON Document stored using the :cas option in set method*

By default the set method runs in synchronous mode. But, the set method can be run in asynchronous mode as follows.

```
client.run do
     client.set("", "") do |ret|
          ret.operation
          ret.success?
          ret.key
          ret.cas
     end
end
```

Asynchronous implies that the method returns immediately and the processing of the script continues while the request is processed on the server and a response returned. In the following modified script the catalog2 key is set using the asynchronous mode. The operation, success, key and cas properties are output.

```
require 'rubygems'
require 'couchbase'
client = Couchbase.connect("http://127.0.0.1:8091")
client.run do
     client.set("catalog2","{'journal': 'Oracle Magazine','publisher': 'Oracle
     Publishing','edition': 'November
December 2013','title': 'Engineering as a Service','author': 'David A. Kelly'}") do |ret|
          print    ret.operation
 print "\n"
          print    ret.success?
print "\n"
          print    ret.key
print "\n"
          print    ret.cas
     end
end
```

When the script is run the operation is output as "set", the success is output as "true", the key is output as "catalog2" and the cas is output as the CAS value of the object stored as shown in Figure 5-18. The cas value would be different for each run of the script.

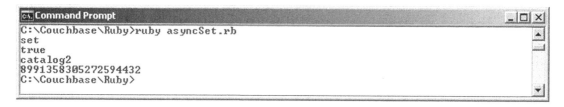

**Figure 5-18.** *Running the asyncSet.rb Script to invoke the set method in asynchronous Mode*

## Adding a Document

The add method is used to create a new key-value pair and cannot be used to set the value of a key already in the server to a new value. The add method has the following signature.

```
object.add(key, value, options)
```

Just as the set method the add method returns the CAS value of the object stored as a fixed number. The method args for the add method are the same as for the set method. Create a Ruby script addDocument.rb. To demonstrate that unlike the set method the add method cannot be used to update a value use the add method on a key that is already defined in the server, the "journal" key for example.

```
require 'rubygems'
require 'couchbase'
client = Couchbase.connect("http://127.0.0.1:8091")
client.add("journal","Oracle Magazine")
```

To run the script invoke the following command.

```
ruby addDocument.rb
```

The script generates the error Couchbase::Error::KeyExists because the journal key is already defined from a previous set method example as shown in Figure 5-19.

```
C:\Ruby21-x64>ruby addDocument.rb
addDocument.rb:7:in `add': failed to store value. The key already exists in the
server. If you have supplied a CAS then the key exists with a CAS value differen
t than specified (key="journal", error=0x0c) (Couchbase::Error::KeyExists)
        from addDocument.rb:7:in `<main>'

C:\Ruby21-x64>_
```

**Figure 5-19.** *Couchbase::Error::KeyExists error*

Delete all the key-value pairs in the Couchbase Server from the Couchbase Admin Console. Deleting key-value pairs using a Ruby script is discussed later in this chapter. Use the add method to add the same key-value pairs as in the following script.

```
require 'rubygems'
require 'couchbase'
client = Couchbase.connect("http://127.0.0.1:8091")
client.add("journal","Oracle Magazine")
client.add("publisher","Oracle Publishing")
client.add("edition","November-December 2013")
client.add("title","Quintessential and Collaborative")
client.add("author","Tom Haunert")
client.add("catalog2","{'journal': 'Oracle Magazine','publisher': 'Oracle
Publishing','edition': 'November December
2013','title': 'Engineering as a Service','author': 'David A. Kelly'}")
```

When the addDocument.rb script is run the key-value pairs in the script get added to Couchbase Server as shown in Figure 5-20.

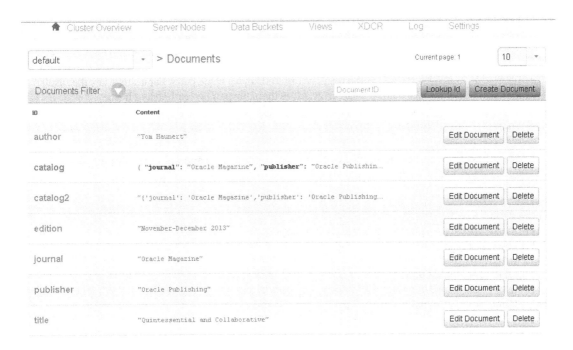

***Figure 5-20.*** *Key/Value Pairs added with the add method*

# Retrieving a Document

The get method is used to retrieve a key-value pair from the server. The get method has the following signature.

```
object.get(keyn [, ruby-get-options ] [, ruby-get-keys ])
```

The get method returns a hash value container with key-value pairs. The method args are discussed in Table 5-2.

***Table 5-2.*** *The get method Arguments*

| Argument | Type | Description |
|---|---|---|
| keyn | String/Symbol/ Array/Hash keyn | Document ID to get. A single key may be specified as a string or, separated multiple key IDs may be specified. ID/IDs may be specified as symbol/symbols such as :journal. ID/IDs may be specified as a hash. |
| ruby-get-options | Hash | Options for the get method. The :format option is used to specify the format of the stored value. The different formats supported are :document for JSON data, :plain for string storage and :marshal to serialize the ruby object using Marshal.dump and Marshal.load. Default format is nil. The :extended option may be used to return the result as ordered key => value pairs. For a single key the result is an array. More than one pair is returned as a hash. |
| ruby-get-keys | Hash | Hash of options containing key-value pairs. |

Create a Ruby script getDocument.rb. Invoke the get method to get and print the value for journal, publisher, edition, title, author, and catalog2.

```
require 'rubygems'
require 'couchbase'
client = Couchbase.connect("http://127.0.0.1:8091")
print client.get("journal")
print "\n"
print client.get("publisher")
print "\n"
print client.get("edition")
print "\n"
print client.get("title")
print "\n"
print client.get("author")
print "\n"
print client.get("catalog2")
```

Run the script with the following command.

```
ruby getDocument.rb
```

The value objects for the specified IDs are returned as shown in Figure 5-21.

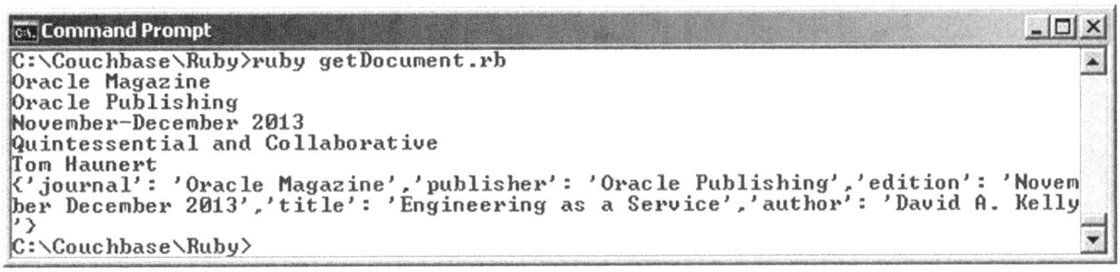

*Figure 5-21.* *Running the getDocument.rb Script to get Key Values*

The default mode for the get method is synchronous, but the get method may be invoked in asynchronous mode with the following notation.

```
client.run do
    client.get("key1", "key2", "key3") do |ret|
      ret.operation
      ret.success?
      ret.key
      ret.value
      ret.flags
      ret.cas
    end
end
```

Create a script asycnGet.rb and invoke the get method on the 'catalog2' key using the asynchronous mode.

```
require 'rubygems'
require 'couchbase'
client = Couchbase.connect("http://127.0.0.1:8091")
client.run do
    client.get("catalog2") do |ret|
      print    ret.operation
 print "\n"
        print    ret.success?
print "\n"
print ret.value
print "\n"
  print    ret.flags
print "\n"
        print    ret.key
print "\n"
        print    ret.cas
    end
end
```

Run the script to get the document value asynchronously. The operation, success, flags, key, and cas are also output as shown in Figure 5-22.

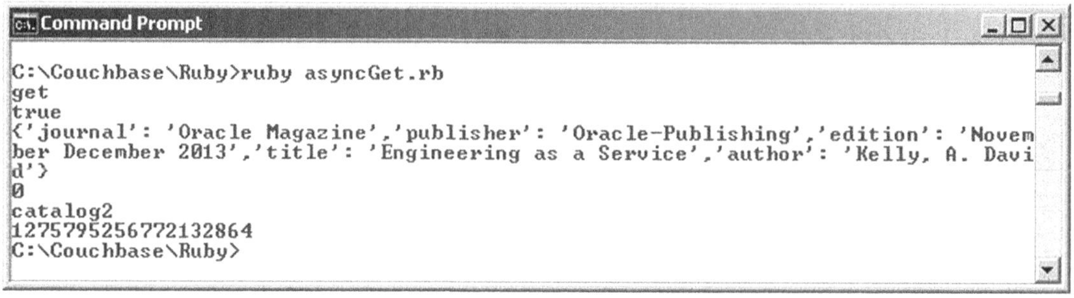

```
C:\Couchbase\Ruby>ruby asyncGet.rb
get
true
{'journal': 'Oracle Magazine','publisher': 'Oracle-Publishing','edition': 'Novem
ber December 2013','title': 'Engineering as a Service','author': 'Kelly, A. Davi
d'}
0
catalog2
1275795256772132864
C:\Couchbase\Ruby>
```

***Figure 5-22.*** *Running the asyncGet.rb Script to get Document Value asynchronously*

The key IDs to be retrieved can also be specified using an array. In the following script an array is created for the IDs to be retrieved and the array is supplied as an argument to the get method.

```
require 'rubygems'
require 'couchbase'
client = Couchbase.connect("http://127.0.0.1:8091")
ary= ["journal","publisher","edition","title","author","catalog2"]
print client.get(ary)
```

The result from the get method is also returned as an array of values in the same order as the document IDs are specified as shown in Figure 5-23.

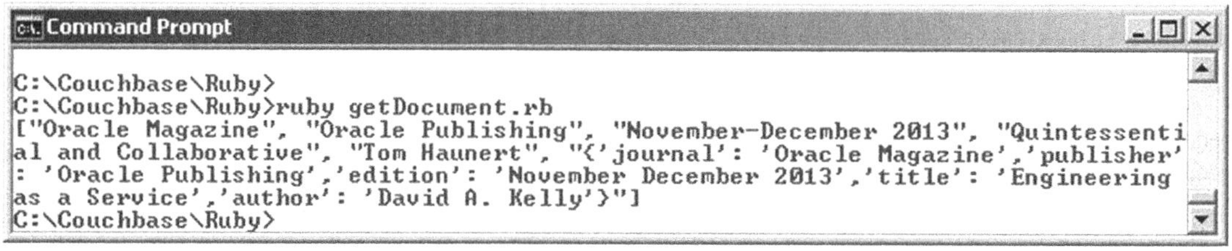

```
C:\Couchbase\Ruby>
C:\Couchbase\Ruby>ruby getDocument.rb
["Oracle Magazine", "Oracle Publishing", "November-December 2013", "Quintessenti
al and Collaborative", "Tom Haunert", "{'journal': 'Oracle Magazine','publisher'
: 'Oracle Publishing','edition': 'November December 2013','title': 'Engineering
as a Service','author': 'David A. Kelly'}"]
C:\Couchbase\Ruby>
```

***Figure 5-23.*** *Getting Documents' Values as an Array*

The document IDs to be retrieved can also be specified as a key/expiry hash. The document is removed after the specified expiry. In the following script the get method arguments are key/expiry hash for the journal and publisher fields.

```
require 'rubygems'
require 'couchbase'
client = Couchbase.connect("http://127.0.0.1:8091")
print client.get("journal" => 10, "publisher" => 20)
```

Run the script to output the JSON document consisting of the key/value pairs for the IDs requested as shown in Figure 5-24.

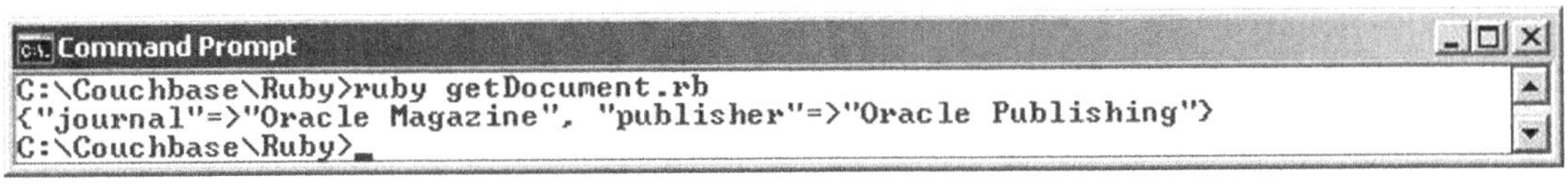

```
Command Prompt                                              _ □ ×
C:\Couchbase\Ruby>ruby getDocument.rb
{"journal"=>"Oracle Magazine", "publisher"=>"Oracle Publishing"}
C:\Couchbase\Ruby>_
```

*Figure 5-24. Specifying get method args as a hash*

The journal and publisher ID documents get removed after running the preceding script. Add the journal and publisher IDs again using the storeDocument.rb script for the next example. Next, we demonstrate the use of the :extended and :format options. In the following script :extended is set to true and :format is set to :document.

```
require 'rubygems'
require 'couchbase'
client = Couchbase.connect("http://127.0.0.1:8091")
print client.get("journal","publisher","edition","title","author","catalog2",:extended =>
true,:format => :document)
```

When the script is run the result is returned as a hash of key=>value, flags, cas pairs. Because the :format is set to :document the result is returned as a JSON document as shown in Figure 5-25.

```
Command Prompt                                              _ □ ×
C:\Couchbase\Ruby>ruby getDocument.rb
{"journal"=>["Oracle Magazine", 0, 5000877085020192768], "publisher"=>["Oracle P
ublishing", 0, 4249416871699941504], "edition"=>["November-December 2013", 0, 141
9677553400283136], "title"=>["Quintessential and Collaborative", 0, 989770715046
0125184], "author"=>["Tom Haunert", 0, 13328607840609566720], "catalog2"=>["{'jo
urnal': 'Oracle Magazine','publisher': 'Oracle Publishing','edition': 'November
December 2013','title': 'Engineering as a Service','author': 'David A. Kelly'}",
 0, 6246137775150923776]}
C:\Couchbase\Ruby>_
```

*Figure 5-25. Using the :extended and :format options in get method*

If the :format is :marshal the key to be retrieved cannot have a value of type JSON document. In the following script the :format is set to :marshal and one of the document IDs is catalog2, which has a JSON document as value.

```
require 'rubygems'
require 'couchbase'
client = Couchbase.connect("http://127.0.0.1:8091")
print client.get("journal","publisher","edition","title","author","catalog2",:extended =>
true,:format => :marshal)
```

When the script is run the Couchbase::Error::ValueFormat error is generated as shown in Figure 5-26.

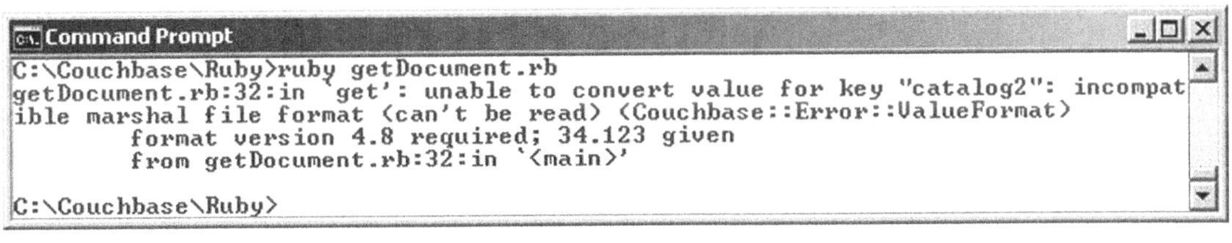

```
Command Prompt                                              _ □ ×
C:\Couchbase\Ruby>ruby getDocument.rb
getDocument.rb:32:in `get': unable to convert value for key "catalog2": incompat
ible marshal file format (can't be read) (Couchbase::Error::ValueFormat)
        format version 4.8 required; 34.123 given
        from getDocument.rb:32:in `<main>'

C:\Couchbase\Ruby>
```

*Figure 5-26. Couchbase::Error::ValueFormat error*

In the following script the :format is set to :plain.

```
require 'rubygems'
require 'couchbase'
client = Couchbase.connect("http://127.0.0.1:8091")
print client.get("journal","publisher","edition","title","author","catalog2",:extended =>
true,:format => :plain)
```

When the script is run the result is in plain format as a string without any conversion as shown in Figure 5-27.

**Figure 5-27.** *Using the :format set to :plain*

If the get method is invoked on key not defined in the Couchbase Server the Couchbase::Error::NotFound error is generated. To demonstrate delete the document with ID catalog2 and run the following script.

```
require 'rubygems'
require 'couchbase'
client = Couchbase.connect("http://127.0.0.1:8091")
print client.get("catalog2")
```

The Couchbase::Error::NotFound error gets generated as shown in Figure 5-28.

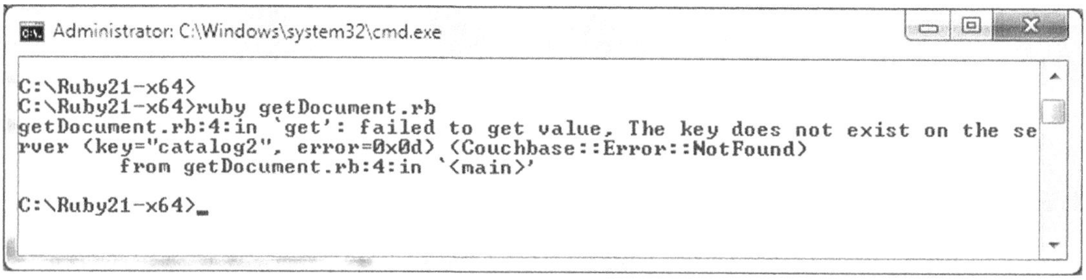

**Figure 5-28.** *Couchbase::Error::NotFound error*

# Updating a Document

The `replace` method may be used to replace a key-value pair. The set method may also be used to update the value of a key already in the server to a new value. The `replace` method has the following signature.

```
object.replace(key, value [, ruby-replace-options ])
```

The `replace` method returns the CAS value of the object stored as a fixed number. The method args are discussed in Table 5-3.

***Table 5-3.*** *The replace method Arguments*

| Argument | Type | Description |
| --- | --- | --- |
| key | string | Document ID used to identify the value. Must be unique in a bucket. |
| value | object | Value to be stored. |
| ruby-replace-options | hash | Options containing key-value pairs for the replace method. |

Create a script `updateDocument.rb`. As an example replace the value of the `catalog2` key with a new JSON object.

```
client.replace("catalog2","{'journal': 'Oracle Magazine','publisher': 'Oracle Publishing',
'edition': 'November December 2013','title': 'Quintessential and Collaborative','author':
'Tom Haunert'}")
```

The `increment` method may be used to increment a numerical value. The `increment` method has the following signature.

```
object.increment(key [, offset ] [, ruby-incr-decr-options ])
```

The `increment` method returns the CAS value of the object stored as a fixed number. The method args are discussed in Table 5-4.

***Table 5-4.*** *The increment method Arguments*

| Argument | Type | Description |
| --- | --- | --- |
| key | string | Document ID used to identify the value. Must be unique in a bucket. |
| offset | Integer | The integer offset value to increment. Default is 1. |
| hash ruby-incr-options | hash | Options containing key-value pairs for the increment method. The :create option may be set a boolean value (true or false) to indicate if the key to be incremented should be created if not already in the server. If the key is created it is initialized to 0 and the value is not incremented in the same operation. The :initial option may be used to initialize a newly created key to a value other than the default 0. The :create option is assumed to be true if the :initial option is specified, regardless of whether the :create option is set or not and regardless of whether the create option is set to false. The :extended option may be used to return an array [value,cas] instead of just value. |

141

The decrement method may be used to decrement a numerical value. The decrement method has the following signature.

```
object.decrement(key [, offset ] [, ruby-incr-decr-options ])
```

The decrement method returns the CAS value of the object stored as a fixed number. The method args are discussed in Table 5-5.

***Table 5-5.*** *The decrement method Arguments*

| Argument | Type | Description |
|---|---|---|
| key | string | Document ID used to identify the value. Must be unique in a bucket. |
| offset | Integer | The integer offset value to decrement. Default is 1. |
| ruby-decr-options | hash | Options containing key-value pairs for the decrement method. The :create option may be set a boolean value (true or false) to indicate if the key to be decremented should be created if not already in the server. If the key is created it is initialized to 0 and the value is not decremented in the same operation. The :initial option may be used to initialize a newly created key to a value other than the default 0. The :create option is assumed to be true if the :initial option is specified, regardless of whether the :create option is set or not and regardless of whether the create option is set to false. The :extended option may be used to return an array [value,cas] instead of just value. |

To demonstrate the increment and decrement methods create numerical key-value pairs; only a numerical value can be incremented/decremented. The key may or may not be numerical.

```
client.set("2",2)
client.set("catalog",1)
```

Increment the value of the "catalog" Id by 2. And, decrement the value of the "2" Id by 1.

```
client.increment("catalog",2)
client.decrement("2",1)
```

Couchbase stores numbers as unsigned values, therefore a value cannot be decremented below 0. If the value of "2" is decremented by 3 instead of 1 the value is decremented to 0.

```
client.decrement("2",3)
```

Negative numbers cannot be incremented using the increment method if after the increment the value is still negative. An integer overflow occurs and a non-logical numerical result is returned. For example, set the key "2" to –3. Subsequently increment the key by 2.

```
require 'rubygems'
require 'couchbase'
client = Couchbase.connect("http://127.0.0.1:8091")
client.set("2",-3)
client.increment("2",2)
```

When the script is run the value is not incremented to –1 as expected but a non-logical numerical value is stored due to an integer overflow as shown in Figure 5-29.

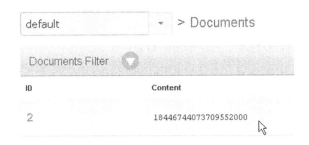

***Figure 5-29.*** *Demonstrating integer overflow*

But if the increment of a negative number makes the value positive the increment is applied as expected.

```
require 'rubygems'
require 'couchbase'
client = Couchbase.connect("http://127.0.0.1:8091")
client.set("2",-1)
client.increment("2",2)
```

If the preceding script is run the value of the "2" key is initially set to –1 and subsequently incremented by 2 to 1 as shown in Figure 5-30.

***Figure 5-30.*** *Demonstrating incrementing a negative number to a positive number*

If the :create option is used to create a new key if not already in the server a new key is created, but not incremented/decremented in the same operation. For example if the server does not have a key "1", the following creates a new key "1" but does not increment the key by 2 in the same operation even though the increment offset is set to 2.

```
client.increment("1",2,:create=>true)
```

To increment the key "1" invoke the increment method again and supply the increment offset.

```
client.increment("1",2)
```

By default a newly created key has the value 0. In the following example the intial value is set to 1 as the :initial option is set to 1. The :create option is not required if the :initial option is set; the :initial option implies that a new key is to be created if not already in the server. If the :create option is set it is ignored. In the following example the :create is set to false, but because :initial is set a new key is created. The :extended is set to true.

```
print client.increment("3",2,:initial=>1,:create=>false, :extended=>true)
```

In the following script updateDocument.rb numerical values are set and incremented/decremented. The replace method is used to replace a JSON document.

```
require 'rubygems'
require 'couchbase'
client = Couchbase.connect("http://127.0.0.1:8091")
client.set("2",2)
client.set("catalog",1)
client.increment("1",2,:create=>true)
client.decrement("2",1)
client.increment("1",2)
print client.increment("3",2,:initial=>1,:create=>false, :extended=>true)
client.increment("catalog",2)
client.replace("catalog2","{'journal': 'Oracle Magazine','publisher': 'Oracle
Publishing','edition': 'November December 2013','title': 'Quintessential and
Collaborative','author': 'Tom Haunert'}")
```

Run the script with the following command.

```
ruby updateDocument.rb
```

One of the increment method invocations has the :extended option set to true. And return value is printed with print. When the script is run the value is initialized to 1 as specified in the :initial option and the CAS value of the object is also returned in the array as shown in Figure 5-31.

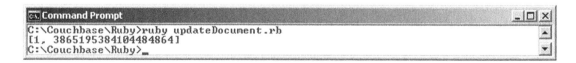

**Figure 5-31.**  *Running the updateDocument.rb Script*

The different key/value pairs in Couchbase Server after running updateDocument.rb Script are as follows as shown in Figure 5-32.

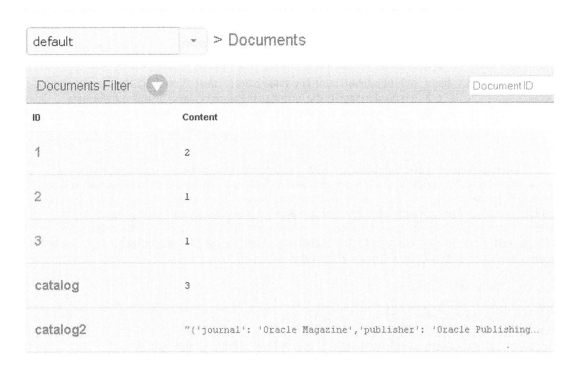

*Figure 5-32.  Key/value pairs in Couchbase Server after running updateDocument.rb*

The replaced `catalog2` document is shown in Couchbase Admin Console in Figure 5-33.

*Figure 5-33.  Replaced catalog2 Document*

The `cas` method may be used to compare and set a value provided the supplied CAS key matches. The `cas` method has the following signature.

```
object.cas(key [, ruby-cas-options ])
```

The `cas` method returns the CAS value of the object stored as a fixed number. The method args are discussed in Table 5-6.

***Table 5-6.*** *The cas method Arguments*

| Argument | Type | Description |
| --- | --- | --- |
| key | string | Document ID used to identify the value. Must be unique in a bucket. |
| ruby-cas-options | hash | Options containing key-value pairs for the cas method. The :format option is used to specify the format of the stored value. The different formats supported are :document for JSON data, :plain for string storage and :marshal to serialize the ruby object using Marshal.dump and Marshal.load. |

The cas method compares the supplied CAS key and provides the value to the subsequent do end block. The value may be updated in the block and the CAS value also gets updated. In the following script the default format is set to :document, implying a JSON document. The "catalog" key is set to a JSON object containing a key/value hash. Subsequently the cas method is invoked with the "catalog" key. The cas method compares and sets/updates the value if the supplied key matches. The CAS value of the object is also updated. The CAS value is output for the original object and the updated object after the cas method is invoked using the :extended option in the get method.

```
require 'rubygems'
require 'couchbase'
client = Couchbase.connect("http://127.0.0.1:8091")
client.default_format = :document
    print ver=   client.set("catalog", {"journal" => "Oracle Magazine"})
         client.cas("catalog") do |val|
                val["publisher"] = "Oracle Publishing"
 val["edition"] = "November December 2013"
 val["title"] = "Engineering as a Service"
 val["author"] = "David A. Kelly"
                val
         end
print client.get("catalog", :extended => true)
```

When the script is run the CAS value of the "catalog" document is output before and after the cas method is invoked. The CAS value is different before and after the cas method is invoked as shown in Figure 5-34.

***Figure 5-34.*** *The CAS value before and after invoking the cas method*

The cas method updates the "catalog" object in the server as shown in Figure 5-35.

*Figure 5-35. The cas method updated catalog key Value*

# Deleting a Document

The delete method may be used to delete a key/value pair. The delete method has the following signature.

```
object.delete(key [, ruby-delete-options ])
```

The delete method returns a boolean (true or false) to indicate if the key got deleted. The method args are discussed in Table 5-7.

*Table 5-7. The delete method Arguments*

| Argument | Type | Description |
|----------|------|-------------|
| key | string | Document ID used to identify the value. Must be unique in a bucket. |
| ruby-delete-options | hash | Options containing key-value pairs for the delete method. The :cas option is used to specify the CAS value for the object for concurrency control. The :quiet option may be set to :false to raise error if the delete method fails in synchronous mode. |

One or more document IDs may be supplied to the delete method. In the following script (deleteDocument.rb), first the "journal" key and the "catalog2" document IDs are deleted individually and subsequently the "publisher" and "edition" IDs are deleted in the same delete method invocation. Add print to the output the value returned by the delete method.

```
require 'rubygems'
require 'couchbase'
client = Couchbase.connect("http://127.0.0.1:8091","default")
print client.delete("journal")
print client.delete("catalog2")
print client.delete("publisher","edition")
```

Add the "journal", "publisher", "edition" and "catalog2" documents prior to running the script. Run the script with the following command.

```
ruby deleteDocument.rb
```

The output indicates that the document IDs get deleted. For multiple document IDs in the same delete method invocation the delete method returns a key=>value hash for the document IDs. A value of true indicates the document gets deleted as shown in Figure 5-36.

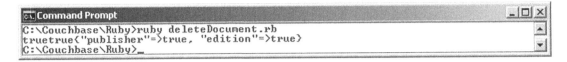

**Figure 5-36.** *Deleting Couchbase Documents*

For concurrency control the `:cas` option may be used to match the CAS value supplied with the CAS value of the object in the server. In the following script the `:cas` option is supplied a value, but not the CAS value returned by the `set` method for the document ID to be deleted.

```
require 'rubygems'
require 'couchbase'
client = Couchbase.connect("http://127.0.0.1:8091","default")
ver = client.set("journal", "Oracle Magazine")
client.delete("journal", :cas => 12345)
```

As the given CAS value does not match the CAS value for the object stored in the server the `Couchbase::Error::KeyExists` error is generated as shown in Figure 5-37.

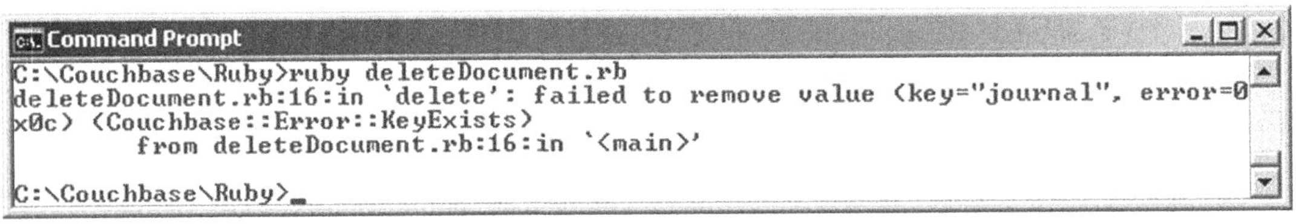

```
C:\Couchbase\Ruby>ruby deleteDocument.rb
deleteDocument.rb:16:in `delete': failed to remove value (key="journal", error=0
x0c) (Couchbase::Error::KeyExists)
        from deleteDocument.rb:16:in `<main>'

C:\Couchbase\Ruby>_
```

**Figure 5-37.** *Couchbase::Error::KeyExists error*

The `:quiet` option is set to true by default, which implies that if the delete fails nil is returned and an error is not raised. In the following script some document IDs that have been deleted are again invoked as arguments to `delete` method. The `:quiet` option is set to false to raise an error if the `delete` method fails. Add the "journal", "publisher", "edition" and "catalog2" documents prior to running the script.

```
require 'rubygems'
require 'couchbase'
client = Couchbase.connect("http://127.0.0.1:8091","default")
print client.delete("journal")
print client.delete("catalog2")
print client.delete("publisher","edition")
print client.delete("journal","publisher","edition",:quiet => false)
```

Run the script, and the `Couchbase::Error::NotFound` error is generated as shown in Figure 5-38.

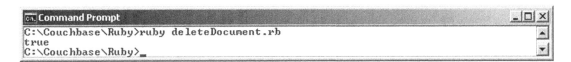

***Figure 5-38.*** *Couchbase::Error::NotFound error*

If the same script is run with :quiet set to true an error is not raised but false is returned for the document IDs that could not be deleted. The reason for not being able to delete is not output as shown in Figure 5-39.

***Figure 5-39.*** *Deleting documents with the :quiet option*

The :cas option was set to a CAS value different than the CAS value stored in the server for the object in an earlier example. In the following script the CAS value supplied to the :cas option is the CAS value stored for the object as returned by the set method.

```
require 'rubygems'
require 'couchbase'
client = Couchbase.connect("http://127.0.0.1:8091","default")
ver = client.set("journal", "Oracle Magazine")
print client.delete("journal", :cas => ver)
```

When the script is run true is returned because the CAS values match and the "journal" document gets deleted as shown in Figure 5-40.

***Figure 5-40.*** *Deleting a document using the :cas option*

# Querying a Document with View

Create a Ruby script queryDocument.rb for querying document/s using a view. Add some documents to Couchbase Server as shown in Figure 5-41.

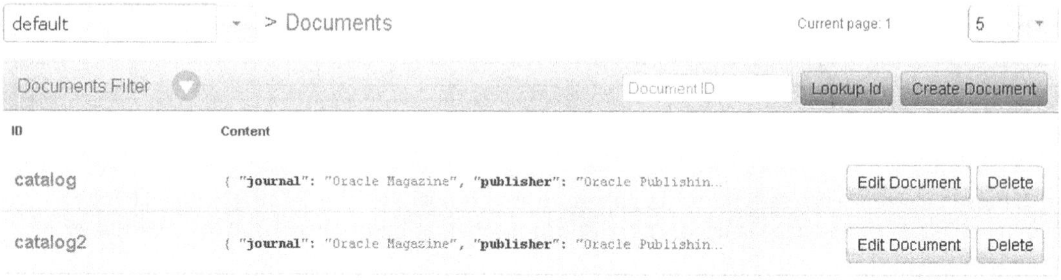

***Figure 5-41.*** *Adding some documents to Couchbase for querying using a view*

Couchbase views were introduced in an earlier chapter. Views are used to query a document stored in Couchabse server. First, we need to create a design document to query, so we'll use the Couchbase Console.

1. Select Views and click on Create Development View.

2. In the Create Development View dialog specify a Design Document Name (dev_catalog for example).

3. Specify a View Name (catalog_view for example). Click on Save.

A design document and a view get added. The view is not usable as such. We need to add map and reduce functions to the view. Map and reduce functions were also introduced in an earlier chapter. Click on the Edit button to edit the catalog_view. The default Map function is listed in the VIEW CODE. Specify the following map function to replace the default map function. The map function emits the document name, journal, publisher, edition, title and author attributes.

```
function (doc, meta) {
  if (meta.type == 'json'){
     emit(doc.name, [doc.journal,doc.publisher,doc.edition,doc.title,doc.author]);
  }
}
```

Click on Save to save the map function. We also need to provide a reduce function. Specify the following function in the Reduce section. The reduce function iterates over the values in the values array and creates a result string to return.

```
function(key, values, rereduce) {
if (!rereduce) {  var result = 0;
for (var i = 0; i < values.length; i++) {
result += values[i];
} return result;
 } else {
return values.length;
}
}
```

Click on the Save button to save the Map and Reduce functions. The catalog_view, including the JSON document on which the view is defined, and the Map and Reduce functions are shown in Figure 5-42.

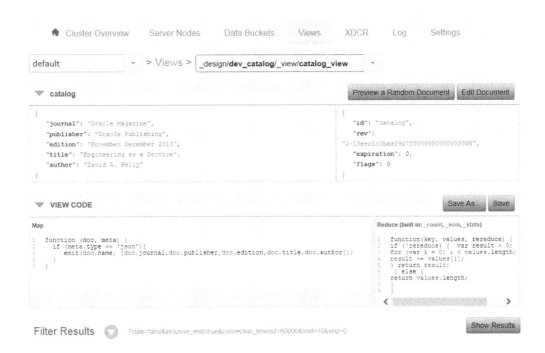

***Figure 5-42.*** *Adding Map and Reduce Functions to a View*

Using the design document 'dev_catalog' and the view 'catalog_view' we shall query the documents in the default bucket. The `design_docs` method in the `Couchbase::Bucket` class is used to obtain the design documents stored in the bucket as a hash. definition for the `dev_catalog` design document. The `dev_catalog` design document is obtained as a `Couchbase::DesignDoc` object from the design documents has as follows.

```
ddoc = client.design_docs["dev_catalog"]
```

The `Couchbase::DesignDoc` class method views is used to obtain the list of views stored in a design document.

```
ddoc.views
```

The `catalog_view` is retrieved as a `Couchbase::View` object from the design document as follows.

```
view=ddoc.catalog_view
```

The each method only yields each document that was fetched by the view as `Couchbase::ViewRow`, which encapsulates a structured JSON document. Because of streaming JSON parser the results are not instantiated by the each method. The results are instantiated when the results are accessed.

Several params may be supplied to the each method, some of which are listed in Table 5-8.

***Table 5-8.*** *The each method Parameters*

| Param | Description |
| --- | --- |
| :include_docs (true, false) | Include full document. Default is true. |
| :quiet (true, false) | Do not raise error if associated document not found. Default is true. |
| :descending | Returns the documents in descending order by key. |
| :key (String, Fixnum, Hash, Array) | Return only documents for the specified key. |
| :limit (Fixnum) | Limit the number of documents in the output. |
| skip (Fixnum) | Skip the specified number of records. |
| :connection_timeout (Fixnum) | Connection timeout for the view request in millisecond. Defaults to 75000. |
| :reduce (true, false) | Whether to use the reduce function. Defaults to true. |
| :stale (String, Symbol, false) | Allow the results from stale view to be used. Valid values are false, :ok and :update_after, the default being :update_after. If set to false view update is forced before returning data. The :ok value allows stale views and the :update_after updates the view after it has been accessed. |
| :group (true, false) | Groups the results using a reduce function to a group or a single row. |

The each method may be used with a block as follows.

```
view.each(:limit => 5, :reduce=>true, :descending=>true) do |row|
   puts row.key
   puts row.value
   puts row.id
   puts row.doc
 end
```

The queryDocument.rb script is listed below.

```
require 'rubygems'
require 'couchbase'
   class Couchbase::Bucket
     alias old_initialize initialize
     def initialize(*args)
       options = args.last
       if options.is_a?(Hash) && options[:environment]
         self.class.send(:define_method, :environment) do
           return options[:environment]
         end
       end
       old_initialize(*args)
     end
   end
```

```
    client = Couchbase::Bucket.new(:environment => :development)
    puts client.design_docs.inspect
ddoc = client.design_docs["dev_catalog"]
print ddoc.views
view=ddoc.catalog_view
  view.each(:limit => 5, :reduce=>true, :descending=>true) do |row|
  puts row.key
  puts row.value
  puts row.id
  puts row.doc
end
```

Run the queryDocument.rb script with the following command.

```
ruby queryDocument.rb
```

The result from the script is shown in the command shell as shown in Figure 5-43.

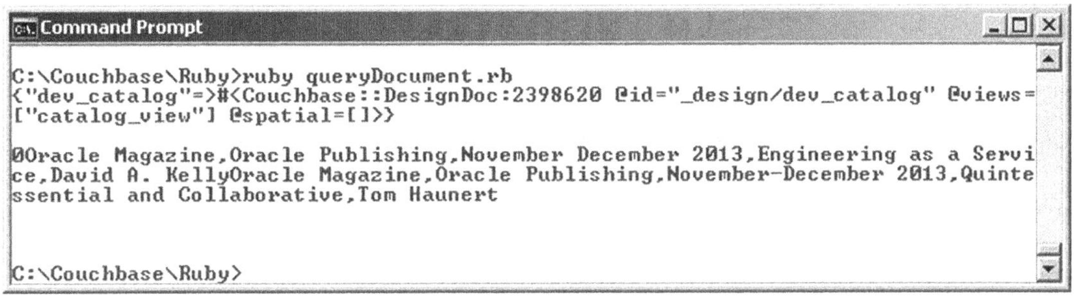

*Figure 5-43.* *Running the queryDocument.rb script to Query Documents*

Alternatively the each method may be invoked to return an Enumerator. Subsequently the attributes of each row returned as a ViewRow object may be output.

```
enum = view.each
enum.map{|row|
puts row.value
puts row.meta
puts row.key
puts row.id
puts row.doc
puts row.data
}
```

The version of queryDocument, which returns a ViewRow, is listed.

```
require 'rubygems'
require 'couchbase'
    class Couchbase::Bucket
      alias old_initialize initialize
      def initialize(*args)
```

```ruby
      options = args.last
      if options.is_a?(Hash) && options[:environment]
        self.class.send(:define_method, :environment) do
          return options[:environment]
        end
      end
      old_initialize(*args)
    end
  end
  client = Couchbase::Bucket.new(:environment => :development)
  puts client.design_docs.inspect
ddoc = client.design_docs["dev_catalog"]
print ddoc.views
view=ddoc.catalog_view
enum = view.each
enum.map{|doc|
puts doc.value
puts doc.meta
puts doc.key
puts doc.id
puts doc.doc
puts doc.data
}
```

The output from the queryDocument.rb script with the Enumerator version of the each method is shown in command shell as you can see in Figure 5-44.

***Figure 5-44.*** *Using the each method and enum to output document values*

# Summary

In this chapter we used a Ruby client library to access Couchbase Server. We added, retrieved, updated and deleted documents using Ruby scripts. We also queried a document using a view. In the next chapter we shall use Node.js to connect to Couchbase and perform similar CRUD operations.

■ ■ ■

# CHAPTER 6

# Using Node.js

Node.js is a lightweight platform built on a V8 JavaScript engine for developing efficient scalable network applications. Node.js is designed for data-intensive real-time applications. Couchbase provides a Node.js Client library for accessing documents stored in Couchbase Server. The Node.js Client library has built-in support for JSON and scales automatically with an expanding Couchbase cluster. In this chapter we shall discuss using Couchbase Server with the Node.js Client library to perform CRUD (Create, Retrieve, Update, Delete) operations on the documents stored in Couchbase. We shall also query Couchbase Server using a View. The chapter covers the following topics.

- Setting Up the Environment
- Connecting with Couchbase Server
- Creating a Document in Couchbase Server
- Getting a Document
- Updating a Document
- Deleting a Document

## Setting Up the Environment

The following software is required to be downloaded and installed for this chapter.

- Couchbase Server
- Node.js
- Node.js Client Library

### Installing Node.js

Download the node-v0.12.0-x64.msi application for Node.js from `http://blog.nodejs.org/2015/02/06/node-v0-12-0-stable/`.

1. Double-click on the msi application to launch the Node.js Setup Wizard.
2. Click on Next in the Setup Wizard as shown in Figure 6-1.

**Figure 6-1.** *Node.js Setup Wizard*

3. Accept the End-User License Agreement and click on Next.

4. In Destination Folder specify a directory to install Node.js in, the default being C:\Program Files\nodejs as shown in Figure 6-2. Click on Next.

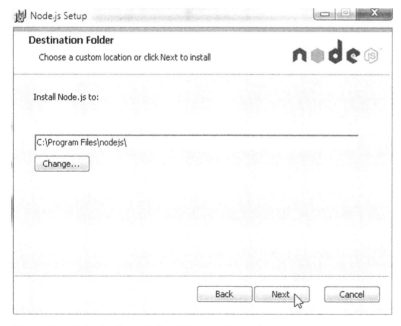

**Figure 6-2.** *Selecting Installation Directory for Node.js*

In Custom Setup, the Node.js features to be installed including the core Node.js runtime are listed for selection as shown in Figure 6-3. Choose the default settings and click on Next.

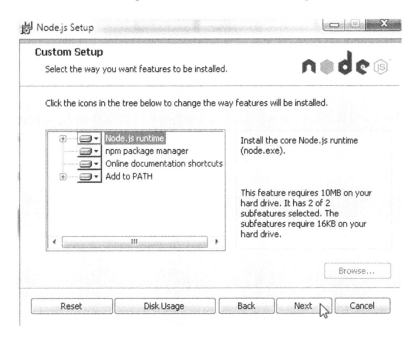

***Figure 6-3.*** *Selecting the Features to Install*

    5.    In Ready to install Node.js, click on Install as shown in Figure 6-4.

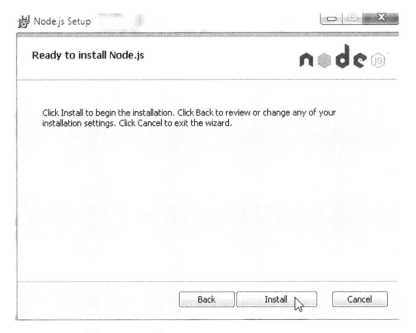

***Figure 6-4.*** *Clicking on Install*

6. The installation of Node.js starts as shown in Figure 6-5. Wait for the installation to finish.

***Figure 6-5.*** *Installing Node.js*

7. When the Node.js completes installing, click on Finish as shown in Figure 6-6.

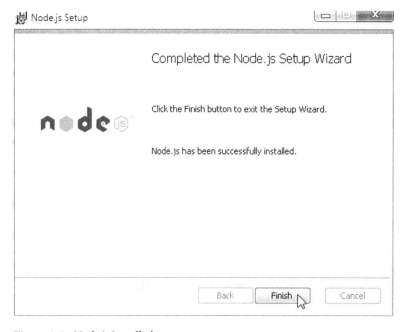

***Figure 6-6.*** *Node.js Installed*

To find the version of Node.js installed, run the following command in a command shell.

```
node --version
```

The output from the command lists the version as 0.12.0 as shown in Figure 6-7.

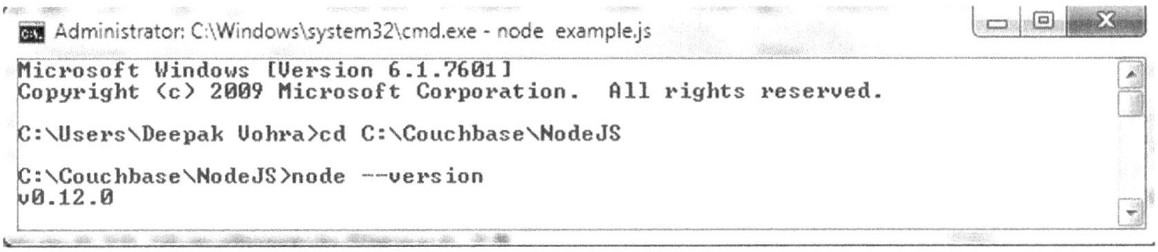

***Figure 6-7.*** *Finding the Node.js Version*

To test the Node.js create a server using the following script; store the script in example.js.

```
var http = require('http');
http.createServer(function (req, res) {
  res.writeHead(200, {'Content-Type': 'text/plain'});
  res.end('Hello World\n');
}).listen(1337, '127.0.0.1');
console.log('Server running at http://127.0.0.1:1337/');
```

From the directory containing the script, run the script with the following command.

```
node example.js
```

The output from the script is shown in the command shell in Figure 6-8.

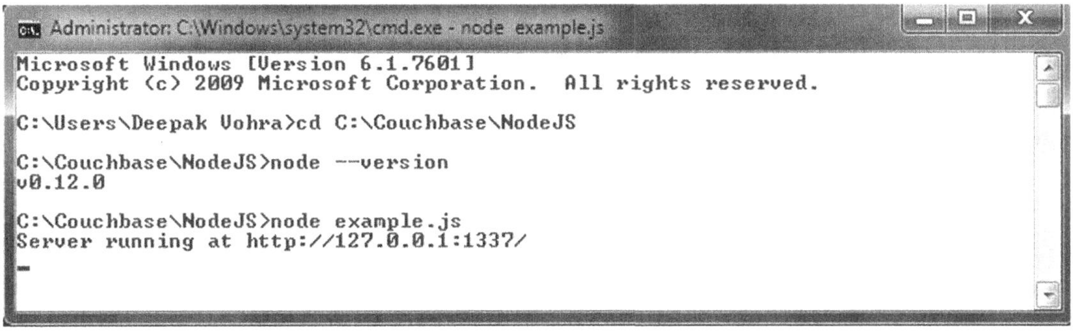

***Figure 6-8.*** *Running the Node.js Example Script*

## Installing Node.js Client Library

Install the Node.js Client Library version 2.08 for Couchbase using npm with the following command in a Windows command shell.

```
npm install couchbase@2.0.8
```

The Node.js Client library gets installed as shown in Figure 6-9.

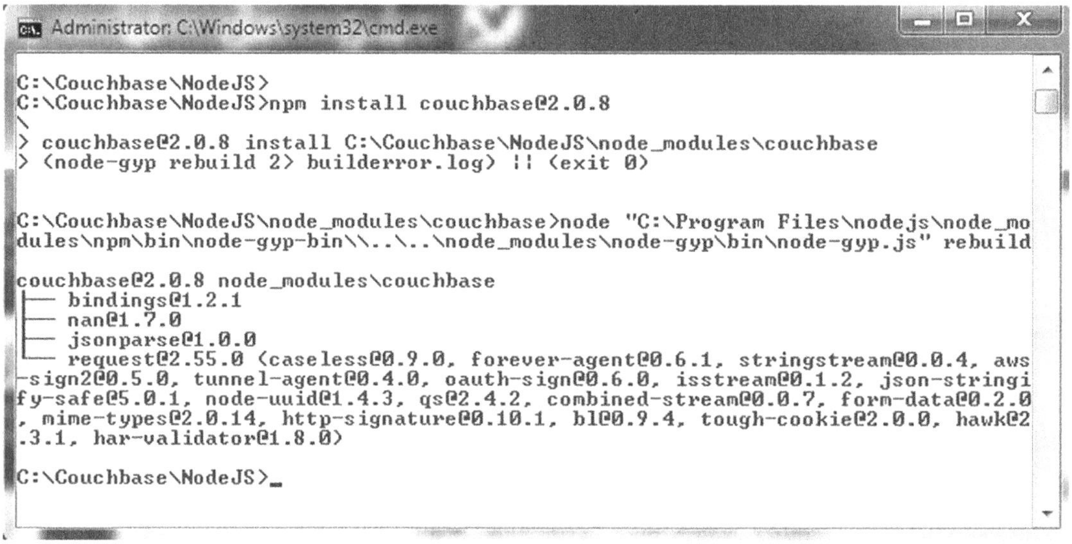

***Figure 6-9.*** *Installing Node.js Client Library for Couchbase*

Next, we shall connect with the Couchbase Server using the Node.js Client library.

# Connecting with Couchbase Server

Create a JavaScript file (connection.js) to connect with the Couchbase Server. First, add a require statement for the Node.js Client library.

```
var couchbase = require('couchbase');
```

The Node.js Client library provides the Cluster class to connect with the Couchbase Server and create a singular cluster containing the buckets. Only one instance of the Cluster class is required in an application. A new Cluster instance can be created using the new operator.

```
new Cluster(cnstr, options)
```

The Cluster class provides a constructor with the parameters listed in Table 6-1.

***Table 6-1.*** *Cluster class Constructor Parameters*

| Parameter | Description |
|-----------|-------------|
| options | Parameter of type Object containing the list of options/properties to be passed to the connection. The only supported option is certpath, the path to the certificate to use for SSL connections. Providing an argument for the parameter is optional. |
| cnstr | Connection string for the cluster. Providing an argument for the parameter is optional. |

Create an instance of Couchbase class using host string as 'localhost:8091'. Alternatively the Ipv4 address may be used instead of localhost. The port is 8091.

```
var cluster = new couchbase.Cluster('couchbase://localhost:8091');
```

The Cluster class provides the openBucket(name, password, callback) method to open a bucket for subsequent operations on the bucket. The openBucket method is asynchronous and returns immediately. Operations on the bucket may be queued subsequent to the method returning. When the connection with the bucket is established the queued operations are run. The openBucket method returns an instance of the Bucket class, which represents a connection with a Couchbase Server bucket. The Bucket class is not to be instantiated directly. Create a callback function that has an err argument for the error generated, if any, by the openBucket method. Output the error, if any, or output a message to indicate that a connection has been established using an if-else statement.

```
callback = function(err)
{
  if (err)
    console.log("Error in establishing connection with Couchbase Server bucket 'default': "+err);
else
  console.log("Connection with Couchbase Server bucket 'default'  established.");
}
```

Invoke the openBucket method using bucket name as 'default' and supply the callback function to invoke when the method returns.

```
var bucket =  cluster.openBucket('default',callback);
```

The openBucket method returns an instance of Bucket class.

The Bucket class provides some class members about the properties of the connection as discussed in Table 6-2.

***Table 6-2.*** *Cluster class Properties*

| Member | Description |
| --- | --- |
| clientVersion | The Node.js client library version as string. |
| configThrottle | Gets/sets the configuration throttling in ms. The bucket waits for the specified ms for a configuration refresh to occur, and if no refresh occurs the bucket forces a configuration refresh. |
| connectionTimeout | The connection timeout in msecs used in the initial connection or re-connection on connection failure. Default value is 5000. |
| lcbVersion | The libcouchbase version. Libcouchbase is the Client library. Node.js builds on the Couchbase C SDK 2.0 version. |
| operationTimeout | The operation timeout in milliseconds. Operation timeout is the time a Bucket waits for a response from the server for a CRUD operation before getting timed out. Default value is 2500. |
| nodeConnectionTimeout | The node connection timeout in milliseconds. Similar to the connection timeout except that it is for a particular node to respond before trying the next node. |
| viewTimeout | The view timeout in milliseconds is the time a Bucket waits for a response from a server for a view request before failing the request with an error. |

Log the Cluster class member settings to the console.

```
console.log("Client Version "+bucket.clientVersion);
console.log("Configuration throttle in msecs "+bucket.configThrottle);
console.log("Connection Timeout in msecs "+ bucket.connectionTimeout);
console.log("Node Connection Timeout msecs "+ bucket.nodeConnectionTimeout);
console.log("libcouchbase version "+bucket.lcbVersion);
console.log("Operation timeout in msecs "+bucket.operationTimeout);
console.log("View timeout in msecs "+bucket.viewTimeout);
```

The complete connection.js is listed below.

```
var couchbase = require('couchbase')
var cluster = new couchbase.Cluster('couchbase://localhost:8091');
callback = function(err)
{
  if (err)
    console.log("Error in establishing connection with Couchbase Server bucket 'default': "+err);
else
  console.log("Connection with Couchbase Server bucket 'default'  established.");
}
var bucket =  cluster.openBucket('default',callback);
console.log("Client Version "+bucket.clientVersion);
console.log("Configuration throttle in msecs "+bucket.configThrottle);
console.log("Connection Timeout in msecs "+ bucket.connectionTimeout);
console.log("Node Connection Timeout msecs "+ bucket.nodeConnectionTimeout);
console.log("libcouchbase version "+bucket.lcbVersion);
console.log("Operation timeout in msecs "+bucket.operationTimeout);
console.log("View timeout in msecs "+bucket.viewTimeout);
```

To connect with Couchbase Server run the following command in a command shell.

```
node connection.js
```

The output from the command indicates that a connection with the server has been established. The connection properties are also output as shown in Figure 6-10.

```
Administrator: C:\Windows\system32\cmd.exe - node connection.js

C:\Couchbase\NodeJS>node connection.js
Client Version 2.0.8
Configuration throttle in msecs 10000000
Connection Timeout in msecs 5000
Node Connection Timeout msecs 2500
libcouchbase version 2.4.9
Operation timeout in msecs 2500
View timeout in msecs 75000
Connection with Couchbase Server bucket 'default'  established.
```

***Figure 6-10.*** *Establishing a Connection with Node.js Server*

## Creating a Document in Couchbase Server

In this section we shall connect with the Couchbase Server and create a JSON document in the server. Two methods are provided in the Bucket class to create a document: upsert() and insert(). The difference between the upsert() and insert() methods is that the upsert() method sets a value for a specified key and overrides the value if the key is already defined in the Couchbase Server, and the insert() method can only be used to add a new key-value pair. If insert() is used with a key already defined in the server, an exception is thrown. Regardless of which method is used for creating a document we need to create a connection with a Couchbase Server bucket as discussed earlier. We also need to specify a key-value pair to be stored in the server. In the following example the key is specified with the catalog_id var, and the value is specified with the catalog var. The value to be stored is a JSON document.

```
var couchbase = require('couchbase')
var cluster = new couchbase.Cluster
('couchbase://localhost:8091');
var bucket =  cluster.openBucket('default');var catalog_id = 'catalog';
var catalog = {
  "journal": "Oracle Magazine",
  "publisher": "Oracle Publishing",
  "edition": "November-December 2013",
  "title": "Quintessential and Collaborative",
  "author": "Tom Haunert"
};
```

## Upserting a Document

Create a JavaScript file storeDocument.js for creating a document using the upsert() method. The connection and key-value pair are specified with the catalog_id and catalog variable, respectively. The upsert() method has the following signature.

```
upsert(key, value, options, callback)
```

The upsert() method parameters are listed in Table 6-3.

***Table 6-3.*** *The upsert() method parameters*

| Parameter | Description |
| --- | --- |
| key | The document key to store of type string or Buffer is a parameter that requires an argument to be supplied. |
| value | The value to store of any type is a parameter that requires an argument to be supplied. |
| options | The options of type Object are optional and discussed in Table 6-4. |
| callback | The callback function is of type Bucket.OpCallback, which has the signature function(error, result). The error parameter is the error thrown by the upsert method and is of type 'undefined' or 'Error'. The result parameter of type Object is the result of the upsert method. |

The options for the upsert() method are discussed in Table 6-4. All options are optional and except the cas option, which does not have a default value, have the default value of 0.

***Table 6-4.*** *The upsert() method Options*

| Option | Type | Description |
| --- | --- | --- |
| cas | Bucket.CAS | Of type CAS, the unique value to use for a document. CAS value is a special object that indicates the state of a document on the server. When the state of a document is modified on the server the CAS value also changes. CAS objects are used by operations that modify value to verify that the value on the server matches – or does not match – a specified CAS value. |
| expiry | number | The initial expiration of the document. The default value of 0 implies that the document does not expire. |
| persist_to | number | Number of nodes to persist operation to. |
| replicate_to | number | Number of nodes to replicate operation to. |

Using the catalog_id as the key and catalog as the value invoke the upsert() method, and in the callback function output the result of the operation to the console.

```
db.upsert(catalog_id, catalog, function(err, result) {
 console.log(result);
 });
```

The storeDocument.js is listed below.

```
var couchbase = require('couchbase')
var cluster = new couchbase.Cluster
('couchbase://localhost:8091');
var bucket =  cluster.openBucket('default');
var catalog_id = 'catalog';
var catalog = {
  "journal": "Oracle Magazine",
  "publisher": "Oracle Publishing",
  "edition": "November-December 2013",
  "title": "Quintessential and Collaborative",
  "author": "Tom Haunert"
};
bucket.upsert(catalog_id, catalog, function(err, result) {
 console.log(result);
 });
```

To store a JSON document using the upsert() method run the following command in a command shell.

```
node storeDocument.js
```

The output from the command is shown in the command shell in Figure 6-11.

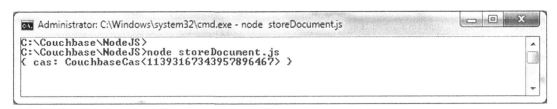

***Figure 6-11.*** *Storing a Document with the storeDocument.js Script*

A document with key as "catalog" gets added to the Couchbase Server as shown in the Couchbase Console in Figure 6-12.

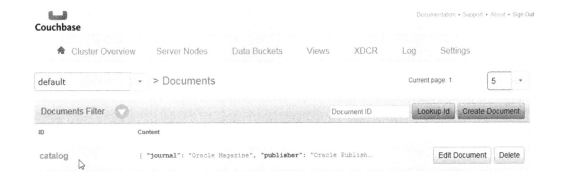

***Figure 6-12.*** *A Document with key catalog in Couchbase Server*

Click on the `catalog` key to display the JSON stored in the document as shown in Figure 6-13.

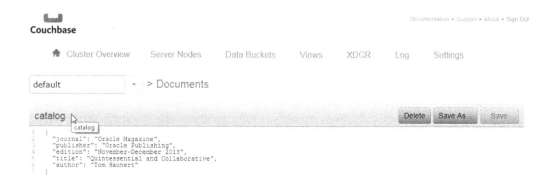

**Figure 6-13.**  *The catalog Document JSON*

## Inserting a Document

Create a JavaScript file `insertDocument.js` for adding a document. As mentioned before, the `insert()` method can only be used to add a new document, implying that the key of the document added must not already be defined in the Couchbase Server. The `insert()` method has the following signature.

```
insert(key, value, options, callback)
```

The `insert` method is identical to the `upsert` method other than the `cas` option not being one of the options and that the method fails if the document with the same key already exists.

To demonstrate that the document key must not be same as another document, create a `insertDocument.js` script and add a document using the insert method and the same key/value pair as used for the upsert method in the previous section. Using the same variables as for the other `storeDocument.js` example, bucket for the Bucket instance, catalog_id for the key, and catalog for the value, add a key-value pair with the `insert` method and output the result in the console in the callback function.

```
bucket.insert(catalog_id, catalog, function(err, result) {
 console.log(result);
 });
```

The `insertDocument.js` JavaScript file is listed below.

```
var couchbase = require('couchbase')
var cluster = new couchbase.Cluster

('couchbase://localhost:8091');

var bucket =  cluster.openBucket('default');
var catalog_id = 'catalog';

var catalog = {
  "journal": "Oracle Magazine",
  "publisher": "Oracle Publishing",
  "edition": "November-December 2013",
```

```
  "title": "Quintessential and Collaborative",
  "author": "Tom Haunert"
};

bucket.insert(catalog_id, catalog, function(err, result) {
 console.log(result);
 });
```

Run the script with the following command.

```
node  insertDocument.js
```

The output from the command is shown in the command shell in Figure 6-14. As indicated by the null value returned, a new document does not get added as a document with the same key-value pair that already exists.

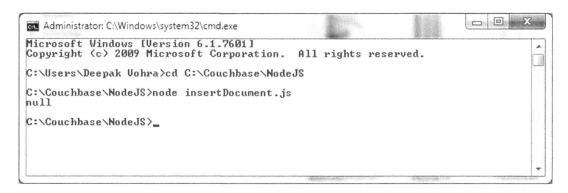

***Figure 6-14.*** *Running insertDocument.js Script does not add a new document*

If the same `insertDocument.js` script is run with a different key, `catalog2` for example, the script returns a new CAS value when run as shown in Figure 6-15.

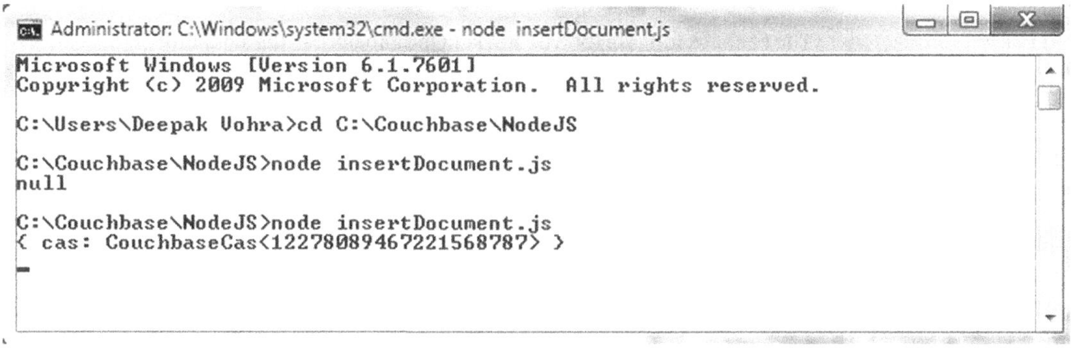

***Figure 6-15.*** *Running insertDocument.js Script to add a Document*

The document with id 'catalog2' gets added to the Couchbase Server as shown in the Couchbase Console in Figure 6-16.

*Figure 6-16.* *Document with id 'catalog2' in Couchbase Server*

# Getting a Document

Create a JavaScript file getDocument.js for getting a document. In this section we shall retrieve a document from Couchbase Server.

The Bucket class provides the get(key, options, callback) method to get a key from a Couchbase cluster/server. The method parameters for the get() method are listed in Table 6-5.

*Table 6-5.* *The get() method Parameters*

| Parameter | Description |
| --- | --- |
| key | The key to get of type string or Buffer is a required parameter. |
| options | The options of type Object are optional. |
| callback | The callback function of type Bucket.OpCallback has the signature function(error, result). The error argument is the error thrown by the get method and is of type 'undefined' or 'Error'. The result parameter of type Object is the result of the get method. |

Using Bucket instance as the bucket variable and key to retrieve as 'catalog', invoke the get() method and output the result retrieved to the console.

```
bucket.get('catalog', function(err, result) {
console.log(result);
});
```

The getDocument.js JavaScript file is listed below.

```
var couchbase = require('couchbase')
var cluster = new couchbase.Cluster('couchbase://localhost:8091');
var bucket = cluster.openBucket('default');
bucket.get('catalog', function(err, result) {
console.log(result);
});
```

To retrieve the document with ID 'catalog', run the following command in a command shell.

```
node getDocument.js
```

The document retrieved is shown in the command shell in Figure 6-17.

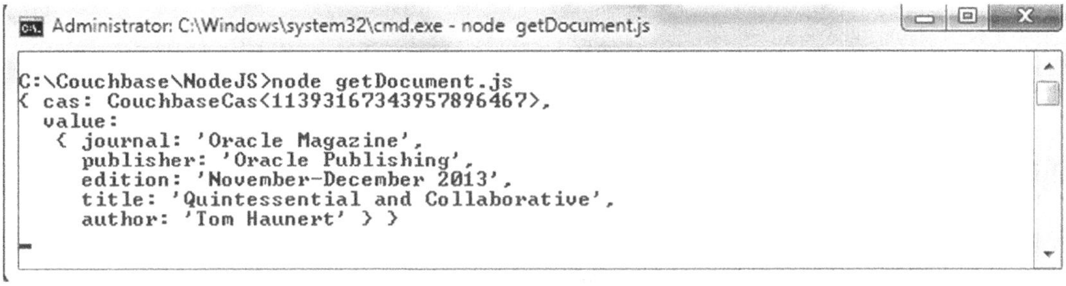

***Figure 6-17.*** *Running the getDocument.js script to retrieve a Document*

The Bucket class provides the getMulti(keys, callback) method to get an array of documents. Create a getMultiDocuments.js script to demonstrate getting multiple documents.

```
var couchbase = require('couchbase')
var cluster = new couchbase.Cluster
('couchbase://localhost:8091');
var bucket = cluster.openBucket('default');
var ids = ['catalog', 'catalog2'];
bucket.getMulti(ids, function(err, results) {
console.log(results);
});
```

The output from getMultiDocuments.js lists the two documents for which keys are specified in the getMulti method as shown in Figure 6-18.

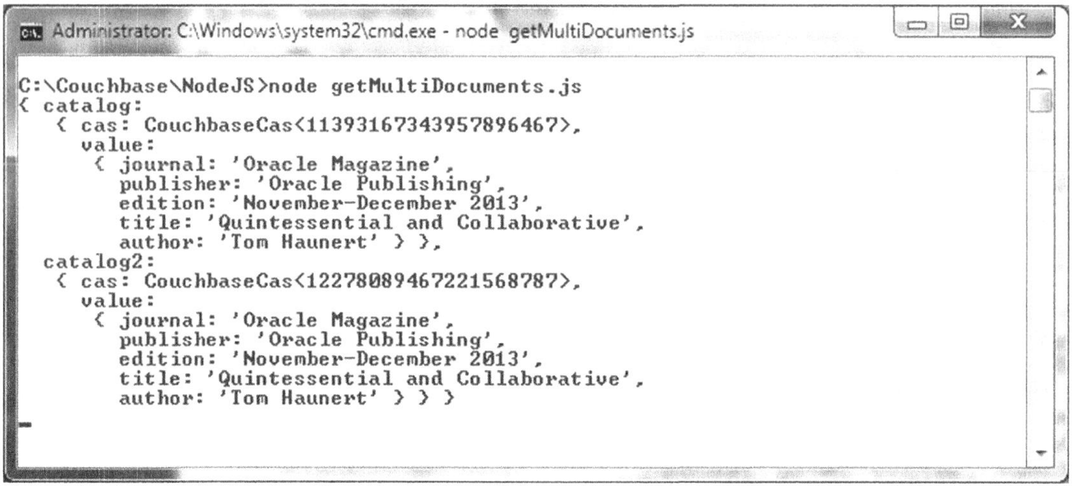

```
Administrator: C:\Windows\system32\cmd.exe - node getMultiDocuments.js

C:\Couchbase\NodeJS>node getMultiDocuments.js
{ catalog:
   { cas: CouchbaseCas<11393167343957896467>,
     value:
      { journal: 'Oracle Magazine',
        publisher: 'Oracle Publishing',
        edition: 'November-December 2013',
        title: 'Quintessential and Collaborative',
        author: 'Tom Haunert' } },
  catalog2:
   { cas: CouchbaseCas<12278089467221568787>,
     value:
      { journal: 'Oracle Magazine',
        publisher: 'Oracle Publishing',
        edition: 'November-December 2013',
        title: 'Quintessential and Collaborative',
        author: 'Tom Haunert' } } }
```

*Figure 6-18.* *Running the getMultiDocuments.js script to to get multiple documents*

# Updating a Document

Create a JavaScript file replaceDocument.js for updating a document. The Bucket class provides the replace() method to replace a document and the method has the following signature.

```
replace(key, value, options, callback)
```

The replace() method has the following parameters listed in Table 6-6.

*Table 6-6.* *The replace() method Parameters*

| Parameter | Description |
|-----------|-------------|
| key | The key to update of type string or Buffer is a parameter for which an argument is required. |
| value | The value to update. |
| options | The options of type object are optional and are the same as for upsert methods as discussed in Table 6-4. |
| callback | The callback function has the signature function(error, result). The error parameter is the error thrown by the replace method and is of type 'undefined' or 'Error.' The result parameter of type object is the result of the replace method. |

We shall update the JSON document with ID 'catalog', which we added earlier. Using Bucket class instance variable bucket, document ID to retrieve as 'catalog', and with a replacement document specified with the catalog2 variable, invoke the replace method. The result is output in the console in the callback function.

```
bucket.replace(catalog_id, catalog2, {}, function(err, result) {
 console.log(result);
 });
```

The replaceDocument.js JavaScript file is listed below.

```
var couchbase = require('couchbase')
var cluster = new couchbase.Cluster
('couchbase://localhost:8091');

var bucket =  cluster.openBucket('default');
var catalog_id = 'catalog';
  var catalog2 =
{
"journal": "Oracle Magazine",
"publisher": "Oracle Publishing",
"edition": "November December 2013",
"title": "Engineering as a Service",
"author": "David A. Kelly",
};
bucket.replace(catalog_id, catalog2, {}, function(err, result) {
 console.log(result);
 });
```

To replace the document with ID 'catalog' run the following command in a command shell.

```
node replaceDocument.js
```

The output from the command is shown in command shell in Figure 6-19.

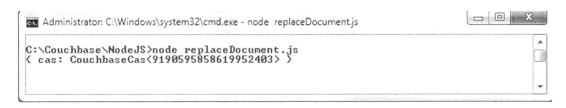

***Figure 6-19.*** *Running the replaceDocument.js to replace a Document*

The updated document is shown in the Couchbase Console in Figure 6-20.

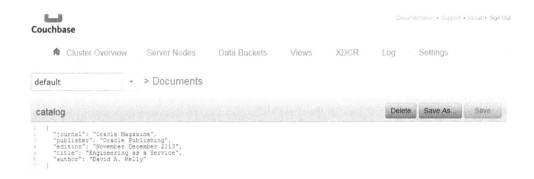

*Figure 6-20.*  *The updated document catalog*

# Deleting a Document

Create a JavaScript file deleteDocument.js for deleting a document. The Bucket class provides the remove()
method to replace a document and the method has the following signature.

```
remove(key, options, callback)
```

The remove() method parameters are discussed in Table 6-7.

*Table 6-7.*  *The remove() method Parameters*

| Parameter | Description |
| --- | --- |
| key | The key to remove of type string or Buffer is a required parameter. |
| options | The options of type Object are optional. All the options are the same as those supported by the upsert method except the expiry option and are discussed in Table 6-4. |
| callback | The callback function has the signature function(error, result). The error parameter is the error thrown by the remove method and is of type 'undefined' or 'Error'. The result parameter of type object is the result of the remove method. |

Next, we shall delete the document with ID 'catalog2'. Use a Bucket class instance to invoke the
remove method.

```
bucket.remove(catalog_id, {},function(err, result) {
console.log(result);
});
```

The deleteDocument.js JavaScript file is listed below.

```
var couchbase = require('couchbase')
var cluster = new couchbase.Cluster
('couchbase://localhost:8091');
var bucket =  cluster.openBucket('default');
var catalog_id = 'catalog2';
bucket.remove(catalog_id, {},function(err, result) {
console.log(result);
});
```

Run the script in a command shell with command node deleteDocument.js as shown in Figure 6-21.

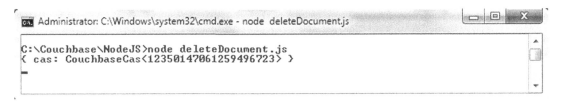

**Figure 6-21.** *Running the deleteDocument.js to delete a Document*

The document with ID 'catalog2' gets removed as shown in the Couchbase Console in Figure 6-22.

**Figure 6-22.** *The document catalog2 is removed and only the document catalog is listed*

The Bucket class provides some other methods for bucket operations such as appending data, prepending data, enabling N1QL, querying using a View query or N1QL query, and incrementing/ decrementing a key's value.

# Summary

In this chapter we discussed the Node.js client library for Couchbase access and Couchbase Server. We also performed CRUD (create, read, update, delete) operations on the Couchbase Server using Node.js scripts. In the next chapter we shall discuss using Elasticsearch with Couchbase.

# CHAPTER 7

# Using Elasticsearch

Elasticsearch is a real-time, full-text search and analytics engine based on RESTful API using JSON over HTTP. Some of the features of Elasticsearch are distributed, highly available, document-oriented, and schema-free. Couchbase Plugin for Elasticsearch makes it feasible to index and search data stored in Couchbase Server in real-time using Elasticsearch. With the plugin, data streams from Couchbase Server to Elasticsearch in real-time. Couchbase data gets indexed in Elasticsearch and may be queried using a RESTful API. To make use of the plugin, two clusters are required to be created: an Elasticsearch cluster and a Couchbase cluster. The plugin is installed in the Elasticsearch cluster. Using the Cross Datacenter Replication (XDCR) in Couchbase Server, the Couchbase data is replicated and streamed to Elasticsearch cluster. The Elasticsearch cluster may be queried to get results as document IDs. The document IDs may be used to retrieve the document from Couchbase Server directly. The Couchbase Server data is kept in sync with the Elasticsearch cluster index. Any changes in the Couchbase data are streamed in real-time to Elasticsearch.

This chapter covers the following topics:

- Setting the Environment
- Installing the Couchbase Plugin for Elasticsearch
- Configuring Elasticsearch
- Installing the Elasticsearch head third-party Plugin
- Starting Elasticsearch
- Providing an index Template in Elasticsearch
- Creating an empty index in Elasticsearch
- Setting the limit on concurrent Requests in Elasticsearch
- Setting the limit on Concurrent Replications in Couchbase Server
- Creating an Elasticsearch Cluster Reference in Couchbase
- Creating a Replication and starting data Transfer
- Querying Elasticsearch
- Adding Documents to Couchbase Server while Replicating
- The Document Count in Elasticsearch

# Setting the Environment

Only specific versions of Elasticsearch Plugin are compatible with specific versions of Elasticsearch. The compatibility matrix for Elasticsearch Plugin/Elasticsearch is listed in Table 7-1.

***Table 7-1.*** *Compatibility Matrix for Elasticsearch Plugin and Elasticsearch*

| Elasticsearch Plugin Version | Elasticsearch Version |
|---|---|
| Elasticsearch Plugin 2.0.0 | 1.3.0 Compatible |
| Elasticsearch Plugin 1.3.0 | 1.0.1 Compatible |
| Elasticsearch Plugin 1.2.0 | 0.90.5 Compatible |
| Elasticsearch Plugin 1.1.0 | 0.90.2 Compatible |
| Elasticsearch Plugin 1.0.0 | 0.19.9 Compatible |

We need to download the following software (different versions may be used if compatible according to Table 7-1):

- Couchbase Server Enterprise Edition 3.0.x from `http://www.couchbase.com/download`.

- Elasticsearch 1.3.0 elasticsearch-1.3.0.zip file from `https://www.elastic.co/downloads/elasticsearch`.

- Curl from `http://curl.haxx.se/download.html`.

Here are the steps to set up your environment:

1. Install the Couchbase Server if not already installed.

2. Double-click on the curl installer .exe file to install curl.

3. Extract the Elasticsearch zip to a directory.

4. Add Elasticsearch bin directory(`C:\elasticsearch\elasticsearch-1.3.0\bin`) and Curl bin directory(`C:\Program Files\cURL\bin`) to the PATH environment variable.

Now we'll create sample data in Couchbase Server. The sample data shall be indexed for search in Elasticsearch cluster. For example, create a document with ID "catalog" and data as the following JSON document.

```
{
  "journal": "OracleMagazine",
  "publisher": "OraclePublishing",
  "edition": "NovemberDecember2013",
  "title": "EngineeringasaService",
  "author": "DavidA.Kelly"
}
```

Add another document with id catalog2.

```
{
  "journal": "OracleMagazine",
  "publisher": "OraclePublishing",
  "edition": "NovemberDecember2013",
  "title": "QuintessentialandCollaborative",
  "author": "TomHaunert"
}
```

The documents in Couchbase are listed in Administration Console as shown in Figure 7-1.

*Figure 7-1.* *Listing Couchbase Documents*

# Installing the Couchbase Plugin for Elasticsearch

Change directory (cd) to the directory in which Elasticsearch is installed, the C:\elasticsearch\elasticsearch-1.3.0 directory for example. Run the following command to install Couchbase Plugin 2.0.0 for Elasticsearch.

```
plugin -install transport-couchbase -url http://packages.couchbase.com.s3.amazonaws.com/
releases/elastic-search-adapter/2.0.0/elasticsearch-transport-couchbase-2.0.0.zip
```

As shown in the output from the command, the plugin gets installed as shown in Figure 7-2.

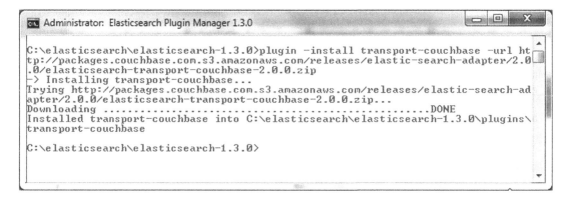

*Figure 7-2.* *Installing Couchbase Plugin for Elasticsearch*

The plugin is not ready for use yet; we need to configure Elasticsearch, such as setting username and password. We also need to install another plugin for Elasticsearch web user interface.

## Configuring Elasticsearch

Set the username and password for the plugin using the following commands.

```
echo couchbase.password: couchbase >> config/elasticsearch.yml
echo couchbase.username: Administrator >> config/elasticsearch.yml
```

The commands update the `C:\elasticsearch\elasticsearch-1.3.0\config/elasticsearch.yaml` configuration file with the username and password. The following two lines get added at the bottom of the elasticsearch.yaml file.

```
couchbase.password: couchbase
couchbase.username: Administrator
```

The output from the commands is shown in Figure 7-3.

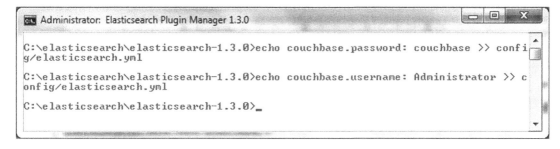

*Figure 7-3.* *Configuring Elasticsearch Username and Password*

The elasticsearch paths may be optionally configured using the configuration settings in `elasticsearch.yaml` listed in Table 7-2.

***Table 7-2.*** *Configuration Settings*

| Configuration Setting | Description | Default setting |
|---|---|---|
| path.data | Path to directory where to store index data allocated for this node. | C:\elasticsearch\ elasticsearch-1.3.0\data |
| path.work | Path to temporary files | |
| path.logs | Path to log files | C:\elasticsearch\ elasticsearch-1.3.0\logs |
| path.plugins | Path to where plugins are installed. The Couchbase plugin for elasticsearch gets installed in this directory. | C:\elasticsearch\ elasticsearch-1.3.0\plugins |

The default configuration settings are used for the parameters not explicitly configured.

# Installing the Elasticsearch Head Third-Party Plugin

We also need to install a third-party plugin for Elasticsearch that provides a web user interface to Elasticsearch called elasticsearch-head. Install the head plugin with the following command.

```
plugin -install mobz/elasticsearch-head
```

The head plugin gets installed in the plugins sub-directory of the Elasticsearch installation. The plugins directory should have two folders called "transport-couchbase" and "head" for the two plugins installed. The output from the command is shown in the command line shell in Figure 7-4.

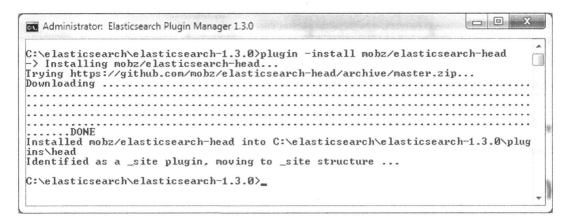

***Figure 7-4.*** *Installing Elasticsearch Head Third-Party Plugin*

# Starting Elasticsearch

Having installed Elasticsearch, the Couchbase plugin for Elasticsearch and the web user interface plugin for Elasticsearch, next we shall start Elasticsearch. Start Elasticsearch with the following command.

```
elasticsearch
```

Elasticsearch cluster gets started as shown in the output from the command in Figure 7-5.

***Figure 7-5.*** *Starting Elasticsearch Cluster*

Access the administrative client user interface to Elasticsearch with the URL http://localhost:9200/ _plugin/head/ in a browser. The Elasticsearch user interface is shown in the browser in Figure 7-6.

**Figure 7-6.** *Elasticsearch Graphical User Interface*

With the Couchbase Server/cluster running and the Elasticsearch cluster running we need make a few other configurations. We need to set the index templates for Elasticsearch and configure Couchbase to replicate data to the remote Elasticsearch cluster.

# Providing an Index Template in Elasticsearch

The index template defines the scope of indexing and searching and is used to index the Couchbase data in Elasticsearch. The default index template is the `plugins/transport-couchbase/couchbase_template.json` template, which is listed.

```
{
    "template" : "*",
    "order" : 10,
    "mappings" : {
        "couchbaseCheckpoint" : {
            "_source" : {
                "includes" : ["doc.*"]
            },
            "dynamic_templates": [
                {
                    "store_no_index": {
                        "match": "*",
                        "mapping": {
                            "store" : "no",
                            "index" : "no",
                            "include_in_all" : false
                        }
                    }
                }
            ]
```

```
        },
        "_default_" : {
            "_source" : {
                "includes" : ["meta.*"]
            },
            "properties" : {
                "meta" : {
                    "type" : "object",
                    "include_in_all" : false
                }
            }
        }
    }
}
```

Another index template may also be used in addition or instead of the default template. To set the index template run the following curl command from the elasticsearch directory, which also applies to all subsequent curl commands.

```
curl -XPUT http://localhost:9200/_template/couchbase -d @plugins/transport-couchbase/
couchbase_template.json
```

If multiple templates are provided, multiple indexes get generated.

# Creating an Empty Index in Elasticsearch

For each Couchbase cluster data bucket create an empty index. We shall be using only the "default" bucket. To create an empty index for the "default" bucket, run the following command.

```
curl -XPUT http://localhost:9200/default
```

The output from the preceding command and the command to set the index template is shown in the command shell in Figure 7-7.

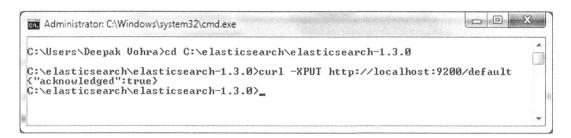

***Figure 7-7.*** *Creating an empty index in Elasticsearch*

An index may be deleted with -XDELETE. For example, the empty index for the default index is deleted as follows.

```
curl -XDELETE  http://localhost:9200/default
```

# Setting the Limit on Concurrent Requests in Elasticsearch

Set the limit on the number of concurrent requests Elasticsearch can handle with the following command.

```
echo couchbase.maxConcurrentRequests: 1024 >> config/elasticsearch.yml
```

Running too many concurrent requests could result in OutOfMemory error. The output from the command is shown in a command shell in Figure 7-8.

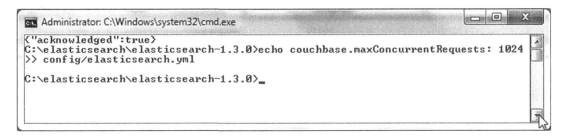

***Figure 7-8.*** *Setting Limit on Concurrent Requests*

For the new configuration settings to take effect we need to restart Elasticsearch. Run the following command to shut down Elasticsearch.

```
curl -XPOST http://localhost:9200/_shutdown
```

Elasticsearch cluster gets shutdown as shown by the output in Figure 7-9.

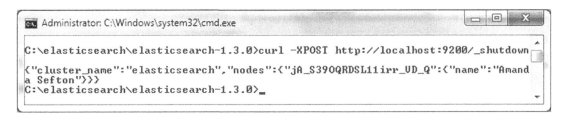

***Figure 7-9.*** *Shutting Down Elasticsearch Cluster*

Restart Elasticsearch with the same command as used before.

```
elasticsearch
```

# Setting the Limit on Concurrent Replications in Couchbase Server

The Elasticsearch nodes have a limit on the indexing data replicated and received from Couchbase Server. To avoid Elasticsearch getting overwhelmed with replication data, run the following command to reduce the limit on concurrent replications from the default value of 32 to 8.

```
curl -X POST -u Administrator:couchbase  http://127.0.0.1:8091/internalSettings  -d
xdcrMaxConcurrentReps=8 --verbose
```

In the preceding command `Administrator:couchbase` are Couchbase admin credentials. The output from the command is shown in a command shell in Figure 7-10.

***Figure 7-10.*** *Setting Limit on Concurrent Replications*

Next, we shall configure Couchbase Server to replicate and send data to Elasticsearch cluster.

# Creating an Elasticsearch Cluster Reference in Couchbase

Couchbase Server provides XDCR to replicate and transfer documents to another cluster. We shall use the XDCR feature to replicate data to the Elasticsearch cluster. XDCR was discussed in Chapter 1. First, we need to configure the Elasticsearch cluster as a remote cluster in Couchbase Server and subsequently we need to start the replication process. The replication runs in real-time and streams all documents in Couchbase Server to Elasticsearch cluster. To create a cluster reference for the Elasticsearch cluster in Couchbase Server, click on XDCR in Couchbase Administration Console in Figure 7-11.

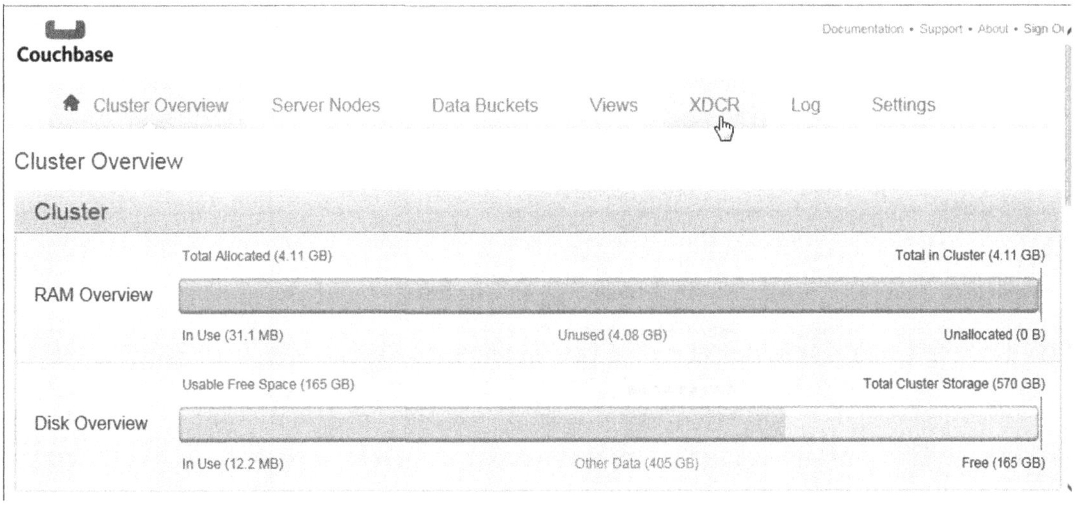

*Figure 7-11.  Selecting XDCR*

Click on Create Cluster Reference button as shown in Figure 7-12.

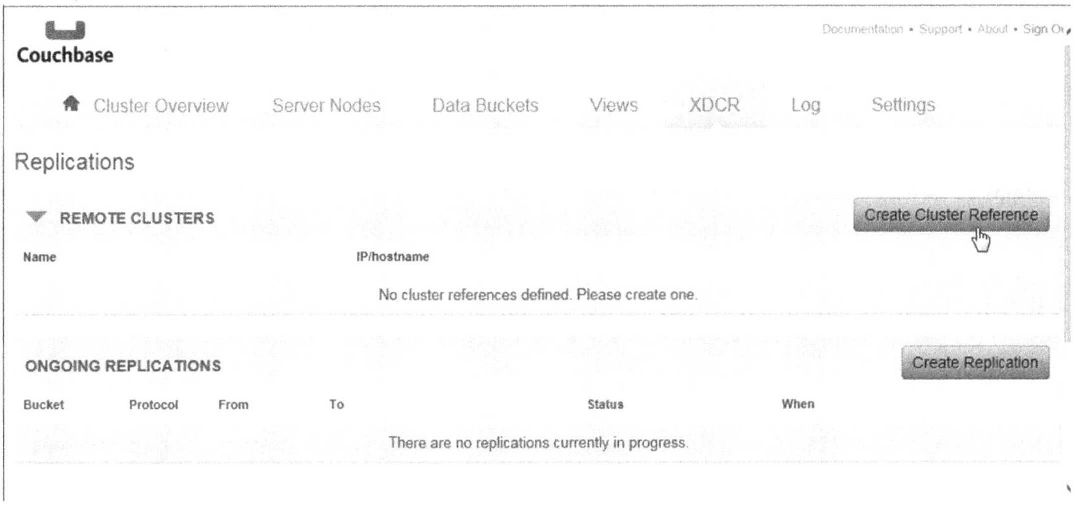

*Figure 7-12.  Selecting Create Cluster Reference*

In the Create Cluster Reference dialog specify a cluster name, which may be obtained from the Elasticsearch web user interface. Specify the IP/hostname, which includes the Ipv4 address of the machine running the Elasticsearch cluster and the port on which Elasticsearch is running. The port is 9091 and the Ipv4 may be obtained with the `ipconfig/all` command. Click on Save as shown in Figure 7-13.

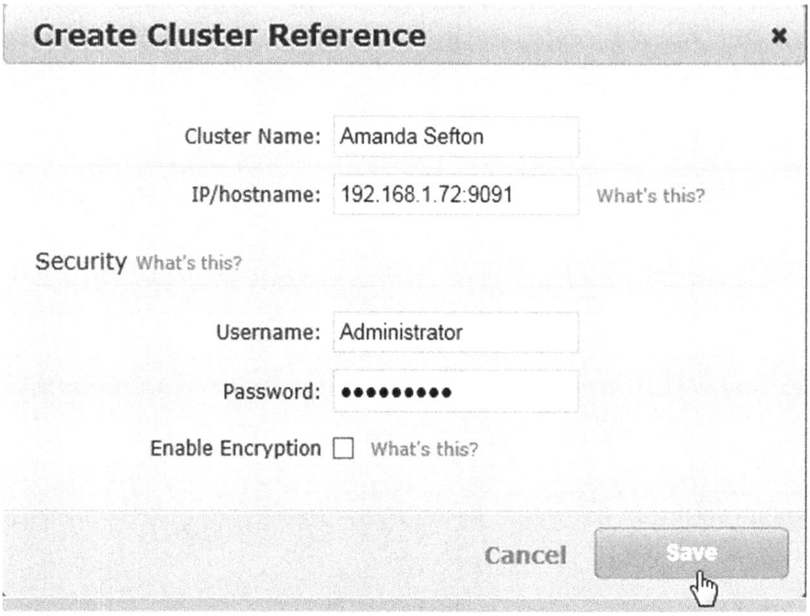

**Figure 7-13.** *Creating Cluster Reference*

A remote cluster reference for the Elasticsearch cluster gets added to Couchbase Administration Console as shown in Figure 7-14.

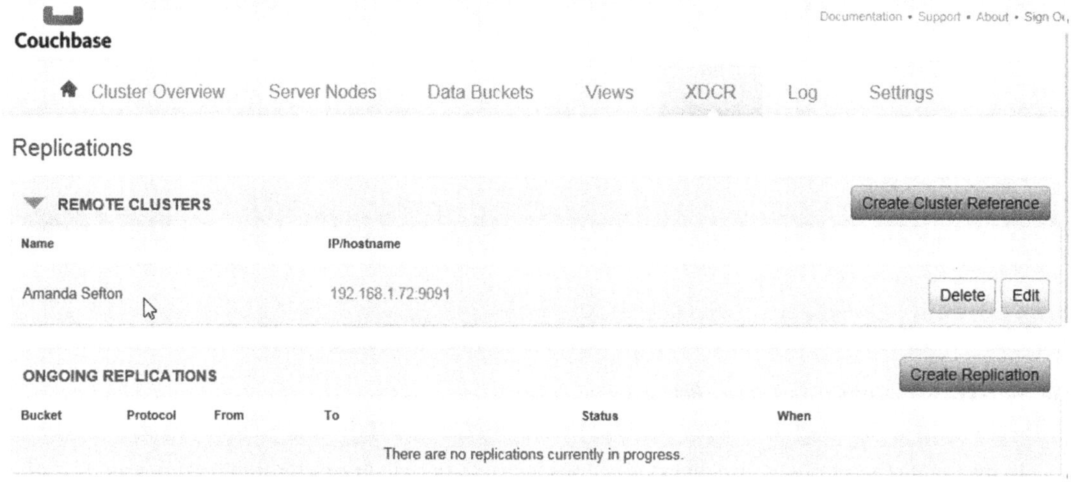

**Figure 7-14.** *Remote Cluster Reference*

# Creating a Replication and Starting Data Transfer

Next, we shall start the replication of the Couchbase Server data to Elasticsearch. Click on Create Replication button as shown in Figure 7-15.

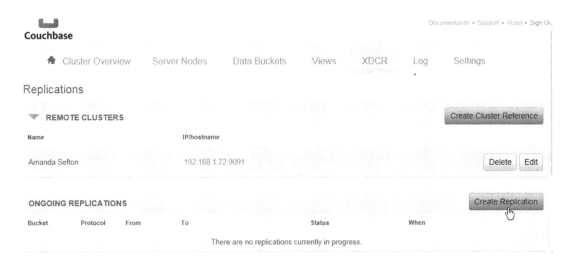

***Figure 7-15.*** *Selecting Create Replication*

In the Create Replication dialog the From cluster is the "this cluster," which is the Couchbase cluster. Select the From data Bucket as the "default" bucket in which we created sample documents. Select the To cluster as Amanda Sefton and the To Bucket as "default." Click on Advanced Settings and select XDCR Protocol as Version 1. Click on Replicate as shown in Figure 7-16.

## Create Replication ✖

Replicate changes from:     To:

     Cluster: **this cluster**          Cluster: | Amanda Sefton    ✔ |

     Bucket: | default      ✔ |          Bucket: | default |

Advanced settings:

| XDCR Protocol: | Version 1    ✔ |
| XDCR Max Replications per Bucket: | 16 |
| XDCR workers per Replication: | 4 |
| XDCR Checkpoint Interval: | 1800 |
| XDCR Batch Count: | 500 |
| XDCR Batch Size (kB): | 2048 |
| XDCR Failure Retry Interval: | 30 |
| XDCR Optimistic Replication Threshold: | 256 |

Cancel     **Replicate**

*Figure 7-16.*  *Creating Replication*

An "ONGOING REPLICATION" for the default bucket gets created from the Couchbase cluster to the Elasticsearch cluster. The Status of the replication is "Starting Up" at first as shown in Figure 7-17.

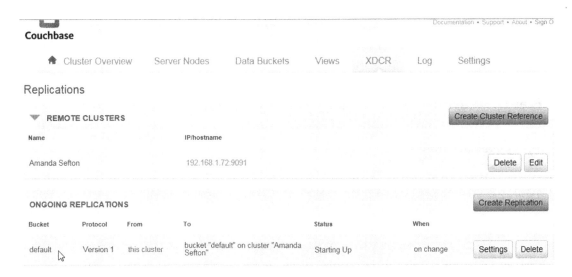

**Figure 7-17.** *"Starting Up" Replication*

When the replication of data begins, the Status changes to "Replicating" as shown in Figure 7-18.

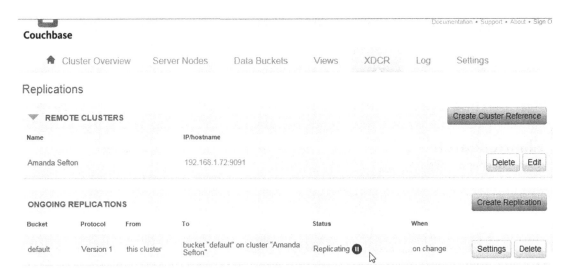

**Figure 7-18.** *"Replicating"*

In the Elasticsearch web user interface, click on Refresh to refresh the status of the indexed documents as shown in Figure 7-19.

189

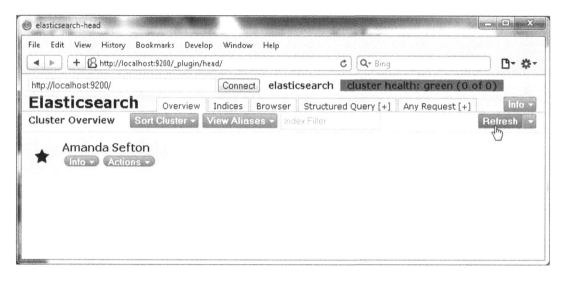

**Figure 7-19.** *Refreshing indexed Document Status*

The docs parameter is the number of documents indexed. The documents in the Couchbase Server are shown indexed in Elasticsearch in Figure 7-20. The number outside the () brackets is the unique number of indexed documents and the number inside the () is the total number of indexed documents across all shards including replicas.

**Figure 7-20.** *Indexed 1027 documents as indicated by docs: 1027*

After the documents have been indexed in Elasticsearch, the documents may be queried using the REST API provided by Elasticsearch.

# Querying Elasticsearch

In this section we shall run some sample queries using the REST API with the curl tool. To search for all JSON documents with query string "NovemberDecember2013" in the document, run the following command in a command shell.

```
curl http://localhost:9200/default/_search?q=NovemberDecember2013
```

The result from the command is displayed. The result is not the actual document but a JSON containing attributes of the document. Some of the document attributes are listed in Table 7-3.

***Table 7-3.*** *Result JSON Attributes*

| Attribute | Description |
|-----------|-------------|
| _index | The index name, which is the same as the bucket name. For the default bucket the index is "default." |
| _type | The type is _couchbaseDocument for the Couchbase document. |
| _id | The id of the document. |
| _source | The metadata of the document. |
| _score | The relevance of the search result to the query as a fraction. A value of 1 being most relevant and a value of 0 being not relevant at all. |

The sample search returns a "total" of two with "_ids" as "catalog" and "catalog2" and with "_index" as "default". By default Elasticsearch indexes each unique word, not phrases. The preceding query for the phrase NovemberDecember2013 returns results for the two JSON documents with the term NovemberDecember2013.

***Figure 7-21.*** *Querying Elasticsearch*

Using an id, a document may be searched in Couchbase Console with the Lookup id button as shown in Figure 7-22.

**Figure 7-22.** *Using Lookup Id to find a document*

Click on Edit Document to display the document as shown in Figure 7-23.

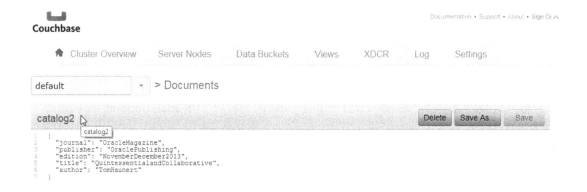

**Figure 7-23.** *Displaying a document*

The query term must not have empty space/s in it. For example, querying search with query Oracle Magazine, which is a phrase, would make use of the following curl command.

```
curl http://localhost:9200/default/_search?q=Oracle Magazine
```

If the publisher field were stored as Oracle Magazine with a space in it only the first word in the query is used for the search. A message "Could not resolve host: Magazine" is returned for the second word as shown in Figure 7-24.

```
C:\Couchbase\elasticsearch-0.90.5>
C:\Couchbase\elasticsearch-0.90.5>curl http://localhost:9200/default/_search?q=O
racle Magazine
{"took":11,"timed_out":false,"_shards":{"total":5,"successful":5,"failed":0},"hi
ts":{"total":1,"max_score":0.35355338,"hits":[{"_index":"default","_type":"couch
baseDocument","_id":"catalog","_score":0.35355338, "_source" : {"meta":{"id":"ca
talog","rev":"2-00000277a8177ed00000000000000000","flags":0,"expiration":0}}]}}
curl: (6) Could not resolve host: Magazine
```

*Figure 7-24. Querying Elasticsearch – another example*

The query string must be provided without quotes. Run the following query as an example.

```
curl http://localhost:9200/default/_search?q="OracleMagazine"
```

An "Empty reply from server" message is received as shown in Figure 7-25.

```
Administrator: Elasticsearch 1.3.0                                    ▭ ▣ ✕

C:\elasticsearch\elasticsearch-1.3.0>
C:\elasticsearch\elasticsearch-1.3.0>curl http://localhost:9200/default/_search?
q="Oracle Magazine"
curl: (52) Empty reply from server

C:\elasticsearch\elasticsearch-1.3.0>_
```

*Figure 7-25. Querying Elasticsearch using double quotes*

The query term must not be a partial term. For example, run a search with query term as "David."

```
curl http://localhost:9200/default/_search?q=David
```

A document id is not returned in the result as shown in Figure 7-26. Instead the search must be run with the full query term as follows.

```
curl http://localhost:9200/default/_search?q=DavidA.Kelly
```

The search result includes a document id as shown in Figure 7-26.

```
Administrator: Elasticsearch 1.3.0                                    ▭ ▣ ✕

C:\elasticsearch\elasticsearch-1.3.0>curl http://localhost:9200/default/_search?
q=David
{"took":2,"timed_out":false,"_shards":{"total":5,"successful":5,"failed":0},"hit
s":{"total":0,"max_score":null,"hits":[]}}
C:\elasticsearch\elasticsearch-1.3.0>curl http://localhost:9200/default/_search?
q=DavidA.Kelly
{"took":3,"timed_out":false,"_shards":{"total":5,"successful":5,"failed":0},"hit
s":{"total":1,"max_score":2.4673123,"hits":[{"_index":"default","_type":"couchba
seDocument","_id":"catalog","_score":2.4673123,"_source":{"meta":{"id":"catalog"
,"rev":"3-13efb792164cdc9b0000000000000000","flags":0,"expiration":0}}]}}
C:\elasticsearch\elasticsearch-1.3.0>_
```

*Figure 7-26. Querying Elasticsearch using a single word-another example*

The Couchbase plugin for Elasticsearch is designed to mainly search JSON documents. Search of document ids is not supported.

# Adding Documents to Couchbase Server while Replicating

Elasticsearch is a real-time search engine and the Couchbase plugin for Elasticsearch replicates and streams Couchbase data store to Elasticsearch cluster in real-time. To demonstrate real-time replication and indexing, add some documents to Couchbase Server while the replication is running as shown in Figure 7-27.

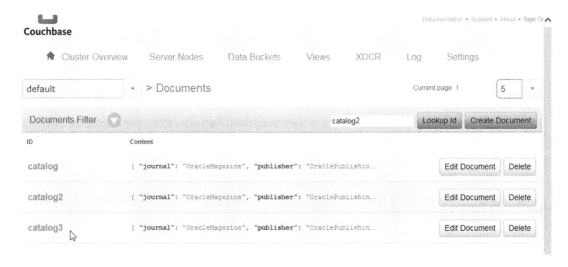

***Figure 7-27.*** *Documents added while the application is running*

The documents added get indexed in real-time. The number of indexed documents as indicated by the docs attribute is shown to have increased with the increase in the number of documents indexed. For example, docs has increased from 1027 to 1028 as shown in Figure 7-28.

elasticsearch-head
File Edit View History Bookmarks Develop Window Help
http://localhost:9200/_plugin/head/
http://localhost:9200/
Connect
elasticsearch
Elasticsearch
Overview
Indices
Browser
Structur
Cluster Overview
Sort Cluster
View Aliases
Index Filter
default
size: 112ki (112ki)
docs: 1028 (1028)
Info
Actions

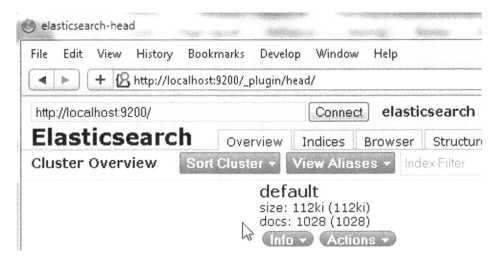

*Figure 7-28.* *Increase in number of indexed documents*

Run the curl query using the REST API again with query string as "OracleMagazine" or "Oracle."

```
curl http://localhost:9200/default/_search?q=OracleMagazine
```

The result includes three document ids including the newly added catalog3 as shown in Figure 7-29.

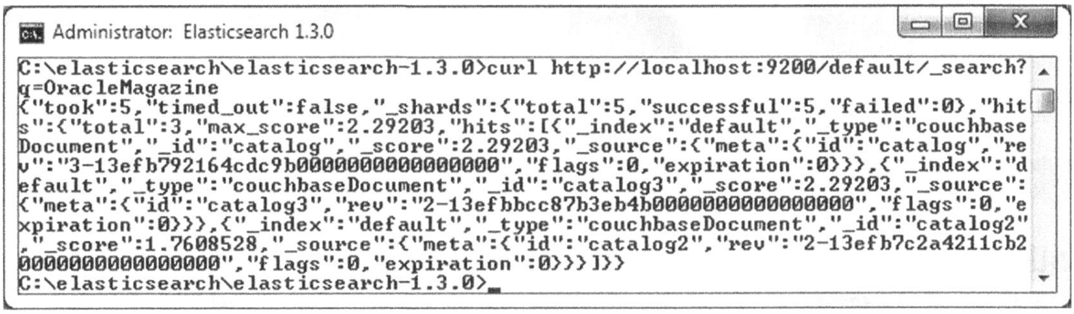

Administrator: Elasticsearch 1.3.0
C:\elasticsearch\elasticsearch-1.3.0>curl http://localhost:9200/default/_search?
q=OracleMagazine
{"took":5,"timed_out":false,"_shards":{"total":5,"successful":5,"failed":0},"hit
s":{"total":3,"max_score":2.29203,"hits":[{"_index":"default","_type":"couchbase
Document","_id":"catalog","_score":2.29203,"_source":{"meta":{"id":"catalog","re
v":"3-13efb792164cdc9b0000000000000000","flags":0,"expiration":0}}},{"_index":"d
efault","_type":"couchbaseDocument","_id":"catalog3","_score":2.29203,"_source":
{"meta":{"id":"catalog3","rev":"2-13efbbcc87b3eb4b0000000000000000","flags":0,"e
xpiration":0}}},{"_index":"default","_type":"couchbaseDocument","_id":"catalog2"
,"_score":1.7608528,"_source":{"meta":{"id":"catalog2","rev":"2-13efb7c2a4211cb2
0000000000000000","flags":0,"expiration":0}}}]}}
C:\elasticsearch\elasticsearch-1.3.0>

*Figure 7-29.* *Elasticsearch query returns multiple documents*

# The Document Count in Elasticsearch

The number of documents added to the index in Elasticsearch is indicated by the docs attribute in the web console for Elasticsearch. The number of documents shown as indexed may be more than the actual number of Couchbase documents indexed. For example, docs is shown as 1027 or 1028 in the preceding examples.

The Item Count in the "default" bucket is listed as 3 as shown in Figure 7-30.

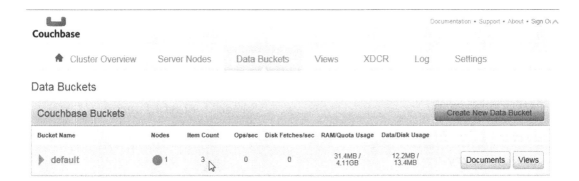

*Figure 7-30.* *Item Count in Couchbase Admin Console*

The difference is because the Couchbase plugin for Elasticsearch and the XDCR sends some additional documents that describe the status of the replication. To get the actual number of documents indexed, run the following curl command.

```
curl http://localhost:9200/default/couchbaseDocument/_count
```

The result has an attribute "count," which is the actual number of documents indexed. A value of 3 is the same as the Item Count in Couchbase console as shown in Figure 7-31.

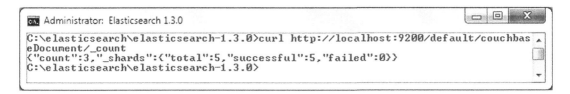

*Figure 7-31.* *Querying Elasticsearch for Document Count*

# Summary

In this chapter we indexed some Couchbase Server documents using the Couchbase Plugin for Elasticsearch. Subsequently we queried the indexed documents. In the next chapter we shall discuss the Couchbase Query Language N1QL.

# CHAPTER 8

## Querying with N1QL

Though Couchbase Server is a NoSQL database it supports an SQL-like query language called N1QL. N1QL supports most of SQL features with additional features suitable for a document-oriented database. N1Ql's use cases include complex queries embedded in applications, and analytics & reporting using ad-hoc queries. Unlike the fixed format of a table in a relational database, the documents stored in Couchbase Server are based on a flexible schema JSON model with nested objects and arrays. The schema-based document model of Couchbase Server requires a flexible path-based language rather than the fixed row/column structure of SQL. N1QL supports different kind of expressions for different kind of operations such as filtering, grouping, and ordering, to mention a few. N1QL queries Couchbase documents, not rows or columns. In this chapter we shall introduce N1QL with some examples:

- Setting the Environment
- Running a SELECT Query
- Filtering with WHERE Clause
- JSON with Nested Objects
- JSON with Nested Arrays
- JSON with Nested Objects and Arrays
- Applying Arithmetic
- Applying ROUND() and TRUNC() Functions
- Concatenating Strings
- Matching Patterns with LIKE & NOT LIKE
- Including and Excluding Null and Missing fields
- Using Multiple Conditions with AND
- Making Multiple Selections with OR Clause
- Ordering Result Set
- Using LIMIT and OFFSET to select a Subset
- Grouping with GROUP BY
- Filtering with HAVING
- Selecting Distinct Values

⬛ **Note**    The N1QL query engine is currently available either as deprecated software (the developer preview of the stand-alone server as described above) or as Beta software (Couchbase 4). To test the N1QL in this chapter, you can choose either option because the N1QL itself will not change, though some features of N1ql could be enhanced. I've chosen to use the DP2 software because that will not change, whereas the Beta software could change between the time of writing and publication.

## Setting Up the Environment

In addition to the Couchbase Server Community Edition (2.x or 3.x) or Enterprise Edition (2.x or 3.x), download the N1QL (Developer Preview 2 or later) couchbase-query_dev_preview2_x86_64_win.zip file from http://cbfs-ext.hq.couchbase.com/tuqtng/. Extract the zip file to a directory and add the directory, for example, C:\Couchbase\N1QL\couchbase-query_dev_preview2_x86_64_win, to the PATH environment variable. To connect to Couchbase Server run the following one of the following commands in a command-line shell.

```
cbq-engine -couchbase http://192.168.1.71:8091
```

or

```
cbq-engine -couchbase http://127.0.0.1:8091
```

The "192.168.1.71" is Ipv4 address and would be different for different users. A connection gets established as shown in Figure 8-1.

*Figure 8-1.* Connecting with Couchbase

To use the command-line query engine run the following command.

```
cbq -engine=http://192.168.1.71:8093
```

or

```
cbq -engine=http://localhost:8093
```

The cbq ➤ prompt gets displayed as shown in Figure 8-2 in which the N1QL queries may be run.

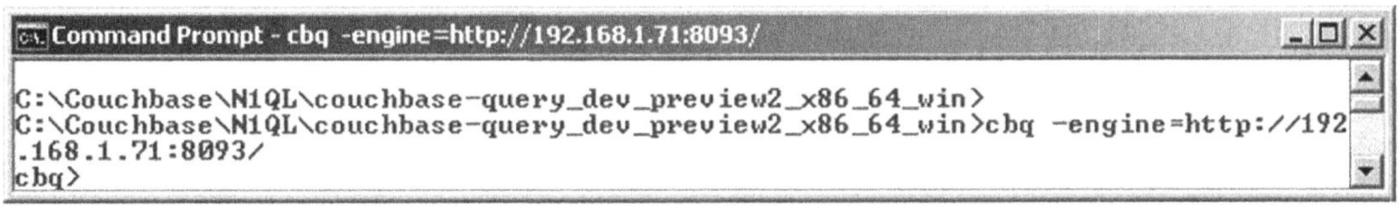

```
Command Prompt - cbq  -engine=http://192.168.1.71:8093/                        _ □ X

C:\Couchbase\N1QL\couchbase-query_dev_preview2_x86_64_win>
C:\Couchbase\N1QL\couchbase-query_dev_preview2_x86_64_win>cbq  -engine=http://192
.168.1.71:8093/
cbq>
```

***Figure 8-2.*** *The cbq ➤ Prompt*

# Running a SELECT Query

The SELECT statement is used to extract data from Couchbase Server. The SELECT statement returns one or more objects as a JSON array result set. To demonstrate the use of SELECT create the following document with id 'catalog' in Couchbase Server using the Couchbase Console.

```
{
  "journal": "Oracle Magazine",
  "publisher": "Oracle Publishing",
  "edition": "November December 2013",
  "title": "Engineering as a Service",
  "author": "David A. Kelly"
}
```

The document is shown in the Couchbase Console in Figure 8-3.

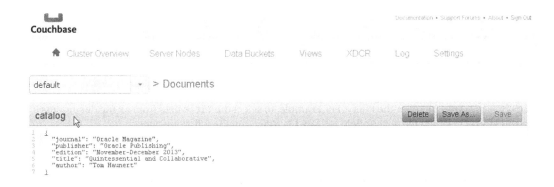

***Figure 8-3.*** *Couchbase Document*

The SELECT statement requires the SELECT clause followed by the result expression, which could be the wildcard * or a document path. The FROM clause specifies the data bucket to query. After adding the preceding document run the following query in the command-line interactive query shell to select all documents from the 'default' bucket.

```
cbq>SELECT * FROM default
```

The result from the query is a JSON document, which contains resultset as the first field. The resultset field is the result set of the N1QL query. The query is run over all the documents in the data bucket. For the example query, the document that we added gets returned as shown in Figure 8-4.

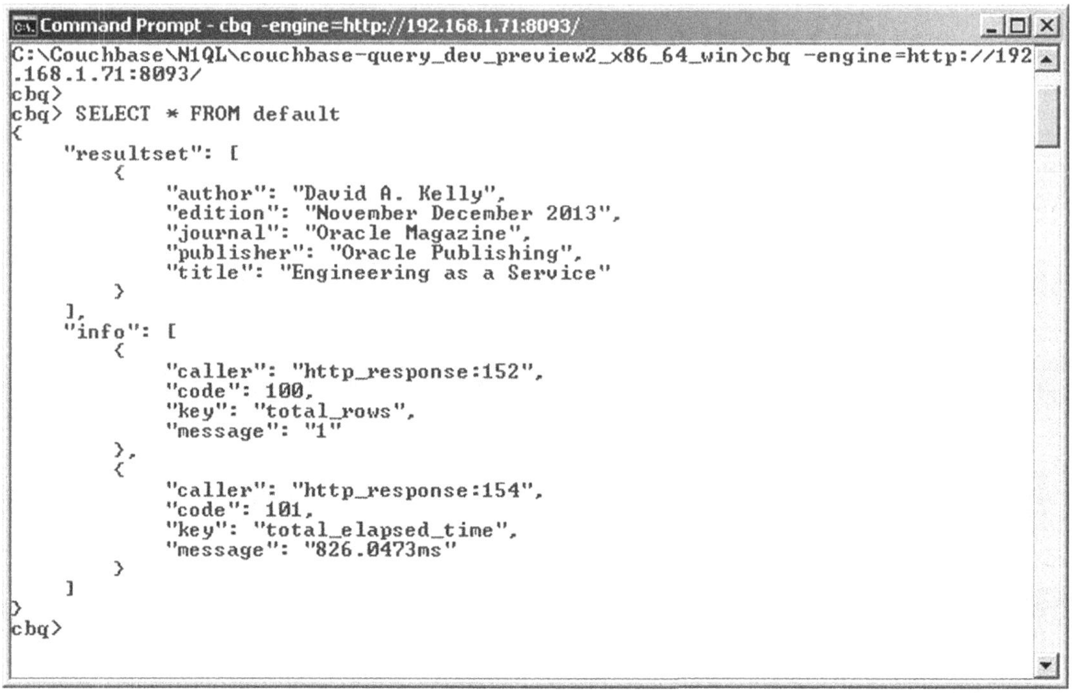

```
Command Prompt - cbq  -engine=http://192.168.1.71:8093/                    _ □ ×
C:\Couchbase\N1QL\couchbase-query_dev_preview2_x86_64_win>cbq  -engine=http://192 ▲
.168.1.71:8093/
cbq>
cbq> SELECT * FROM default
{
    "resultset": [
        {
            "author": "David A. Kelly",
            "edition": "November December 2013",
            "journal": "Oracle Magazine",
            "publisher": "Oracle Publishing",
            "title": "Engineering as a Service"
        }
    ],
    "info": [
        {
            "caller": "http_response:152",
            "code": 100,
            "key": "total_rows",
            "message": "1"
        },
        {
            "caller": "http_response:154",
            "code": 101,
            "key": "total_elapsed_time",
            "message": "826.0473ms"
        }
    ]
}
cbq>                                                                              ▼
```

*Figure 8-4.* *Running a SELECT query on the default bucket*

The JSON document also includes an info field, which has the metadata for the query.

# Filtering with WHERE Clause

The WHERE clause is used to filter the query. For example, to select the title field from all the documents in the default bucket in which the author field is 'David A. Kelly' run the following query.

```
SELECT title FROM default WHERE author='David A. Kelly'
```

The query returns the title as 'Engineering as a Service.'

```
Command Prompt - cbq -engine=http://192.168.1.71:8093/                    _ □ ×
cbq>
cbq> SELECT title FROM default WHERE author='David A. Kelly'
{
    "resultset": [
        {
            "title": "Engineering as a Service"
        }
    ],
    "info": [
        {
            "caller": "http_response:152",
            "code": 100,
            "key": "total_rows",
            "message": "1"
        },
        {
            "caller": "http_response:154",
            "code": 101,
            "key": "total_elapsed_time",
            "message": "256.0146ms"
        }
    ]
}
cbq> _
```

*Figure 8-5.* *Running a SELECT Query with a WHERE clause Filter*

We added only one document to the database bucket, therefore only one title is returned. Add another document with id catalog2.

```
{
  "journal": "Oracle Magazine",
  "publisher": "Oracle Publishing",
  "edition": "November-December 2013",
  "title": "Quintessential and Collaborative",
  "author": "Tom Haunert"
}
```

Couchbase Server has two documents as shown in the Couchbase Admin Console as shown in Figure 8-6.

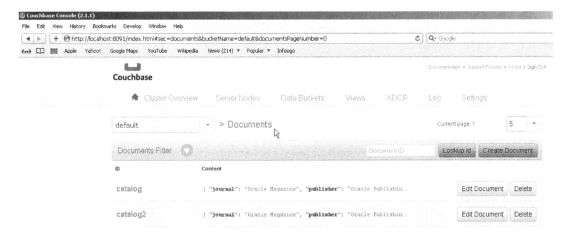

*Figure 8-6.* *Two Documents in Couchbase Server*

Run the same query again.

```
SELECT * FROM default
```

The resultset field contains two JSON documents as shown in Figure 8-7.

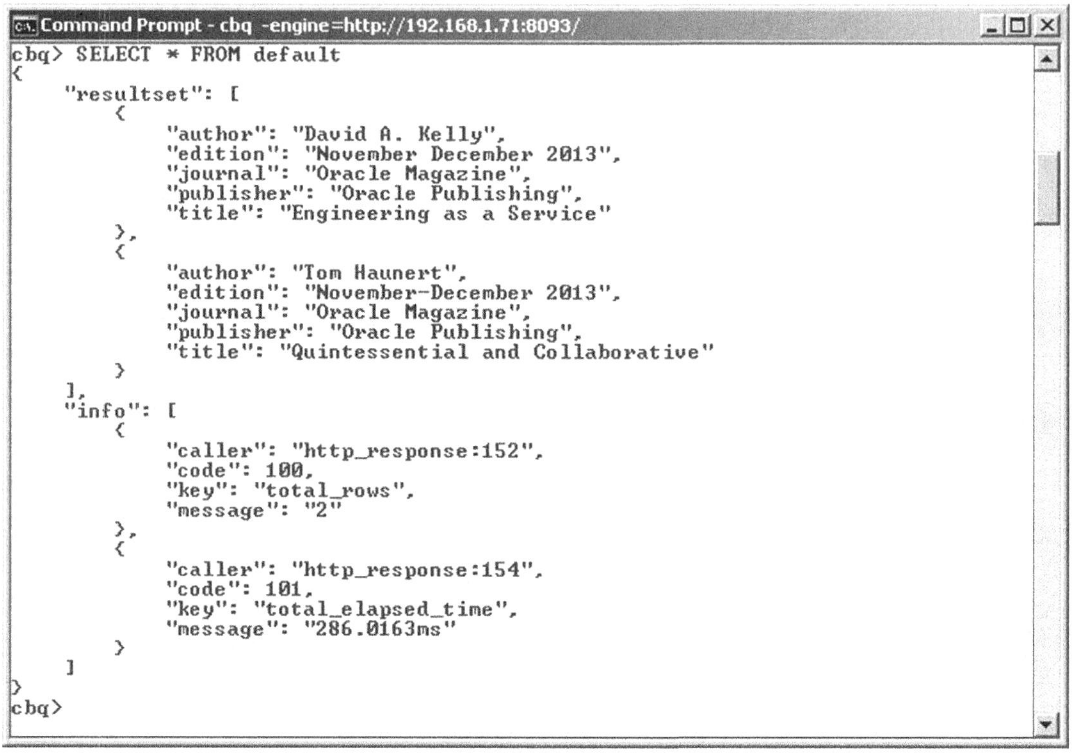

```
Command Prompt - cbq -engine=http://192.168.1.71:8093/                    _ □ ×
cbq> SELECT * FROM default
{
    "resultset": [
        {
            "author": "David A. Kelly",
            "edition": "November December 2013",
            "journal": "Oracle Magazine",
            "publisher": "Oracle Publishing",
            "title": "Engineering as a Service"
        },
        {
            "author": "Tom Haunert",
            "edition": "November-December 2013",
            "journal": "Oracle Magazine",
            "publisher": "Oracle Publishing",
            "title": "Quintessential and Collaborative"
        }
    ],
    "info": [
        {
            "caller": "http_response:152",
            "code": 100,
            "key": "total_rows",
            "message": "2"
        },
        {
            "caller": "http_response:154",
            "code": 101,
            "key": "total_elapsed_time",
            "message": "286.0163ms"
        }
    ]
}
cbq>
```

***Figure 8-7.*** *SELECT Query returns two Documents*

Run the following N1QL query to select all the 'title' fields from the data bucket 'default'.

```
SELECT title FROM default
```

The result set has two titles as compared to only one title when the default bucket had only one document as shown in Figure 8-8.

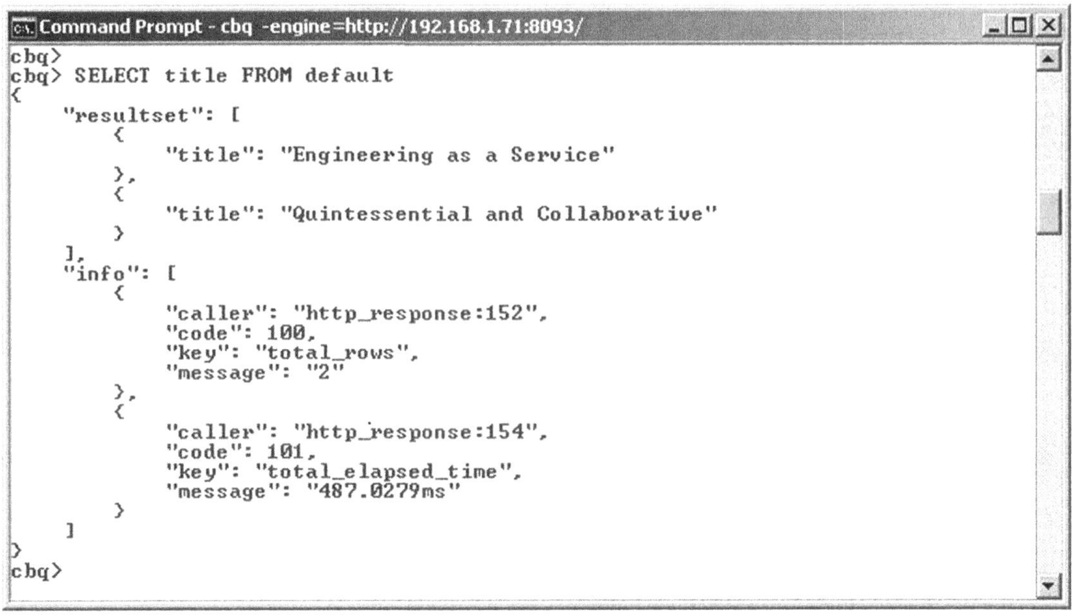

```
Command Prompt - cbq  -engine=http://192.168.1.71:8093/                    _ □ X
cbq>
cbq> SELECT title FROM default
{
    "resultset": [
        {
            "title": "Engineering as a Service"
        },
        {
            "title": "Quintessential and Collaborative"
        }
    ],
    "info": [
        {
            "caller": "http_response:152",
            "code": 100,
            "key": "total_rows",
            "message": "2"
        },
        {
            "caller": "http_response:154",
            "code": 101,
            "key": "total_elapsed_time",
            "message": "487.0279ms"
        }
    ]
}
cbq>
```

***Figure 8-8.*** *Selecting 'title' from the default Bucket*

# JSON with Nested Objects

When referring to nested documents the '.' operator is used to refer to children, and the '[]' is used to refer to an element in an array. You can use a combination of these operators to access data at any depth in a document. The documents added to Couchbase were simple JSON documents without any nested objects or arrays. Next, add the following document with ID catalog4, which has nested JSON objects.

```
{
  "journal": "Oracle Magazine",
  "publisher": "Oracle Publishing",
  "edition": "November-December 2013",
  "title": {
    "title1": "Engineering as a Service",
    "title2": "Quintessential and Collaborative"
  },
  "author": {
    "author1": "David A. Kelly",
    "author2": "Tom Haunert"
  }
}
```

The document is shown in Couchbase Console as shown in Figure 8-9.

**Figure 8-9.** *Couchbase Document with nested JSON Objects*

The '.' operator is used to reference the nested objects. For example, the following query is used to select the title1 field from the nested object title.

```
SELECT title.title1   FROM default
```

The result set returns the title1 field in the nested object title as shown in Figure 8-10.

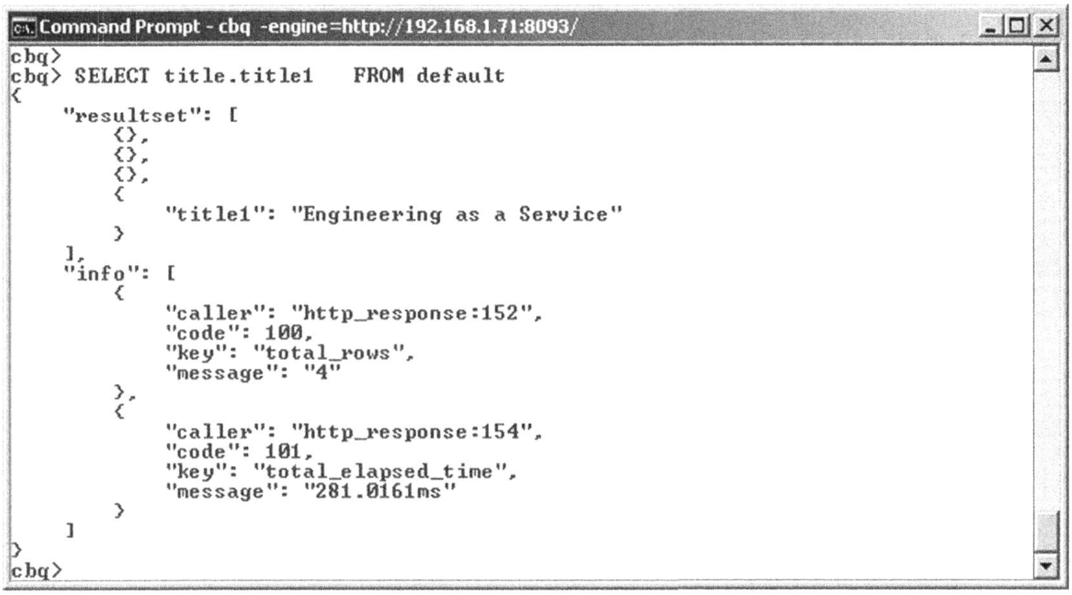

**Figure 8-10.** *Using the '.' Operator to access a nested object*

A field in a nested object may also be referred to using the [ ] operator with the field name in the specified in the [ ]. For example, the following query selects the author1 field in the nested object author.

```
SELECT author["author1"]    FROM default
```

As author1 in author is David A. Kelly, the result set includes the same as shown in Figure 8-11.

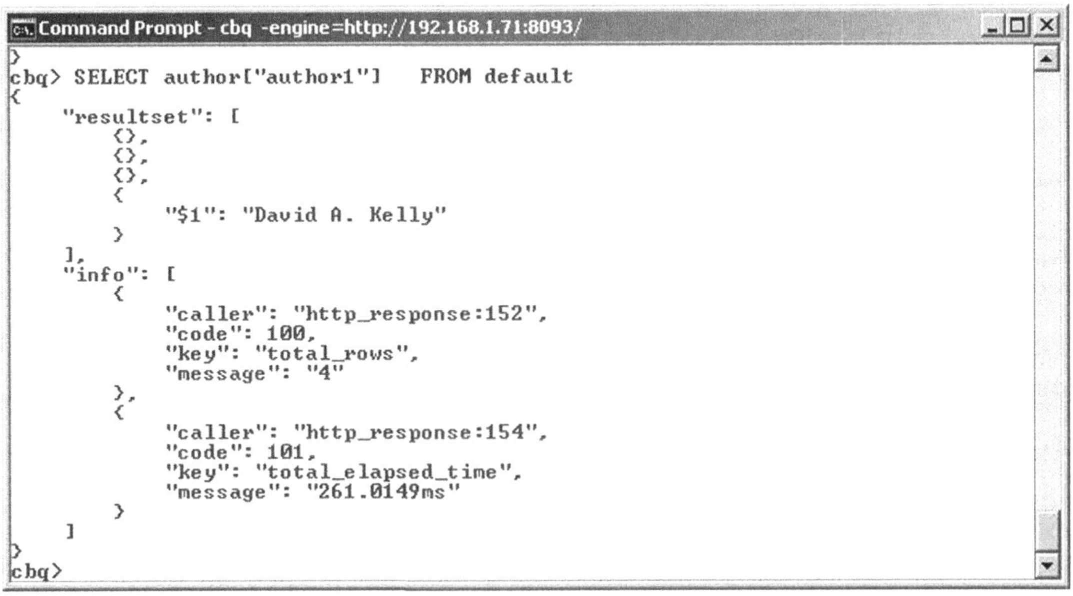

**Figure 8-11.** *Using the [ ] operator to access a field in a nested object*

# JSON with Nested arrays

The preceding example had nested JSON objects, but a top-level JSON document may also have nested arrays. Add a document with ID catalog3 with nested arrays for the title and the author fields.

```
{
  "journal": "Oracle Magazine",
  "publisher": "Oracle Publishing",
  "edition": "November-December 2013",
  "title": [
    "Engineering as a Service",
    "Quintessential and Collaborative"
  ],
  "author": [
    "David A. Kelly",
    "Tom Haunert"
  ]
}
```

The catalog3 document is shown in the Couchbase Console as shown in Figure 8-12.

***Figure 8-12.*** *Couchbase Document with nested Arrays*

To refer an element in an array the [ ] operator is used. Array elements are indexed starting from 0. For example, the following query selects the first element in the title array. The AS clause is used to specify the identifier for the result returned by the query.

```
SELECT title[0] AS title1  FROM default
```

The query returns the first element in the title array and is identified as title1 as shown in Figure 8-13.

```
Command Prompt - cbq  -engine=http://192.168.1.71:8093/                        _ □ ×
cbq> SELECT title[0] AS title1 FROM default
{
    "resultset": [
        {},
        {},
        {
            "title1": "Engineering as a Service"
        }
    ],
    "info": [
        {
            "caller": "http_response:152",
            "code": 100,
            "key": "total_rows",
            "message": "3"
        },
        {
            "caller": "http_response:154",
            "code": 101,
            "key": "total_elapsed_time",
            "message": "277.0158ms"
        }
    ]
}
cbq>
```

***Figure 8-13.*** *Using the [ ] operator to access a nested array element*

The same query may select multiple fields/elements. For example, the following query selects the first element in the title array and the second element in the author array.

```
SELECT title[0], author[1] FROM default
```

The result set includes the selected array elements as shown in Figure 8-14. Because an AS clause is not specified in the query, the resultset fields are identified by the default $1 and $2 identifiers.

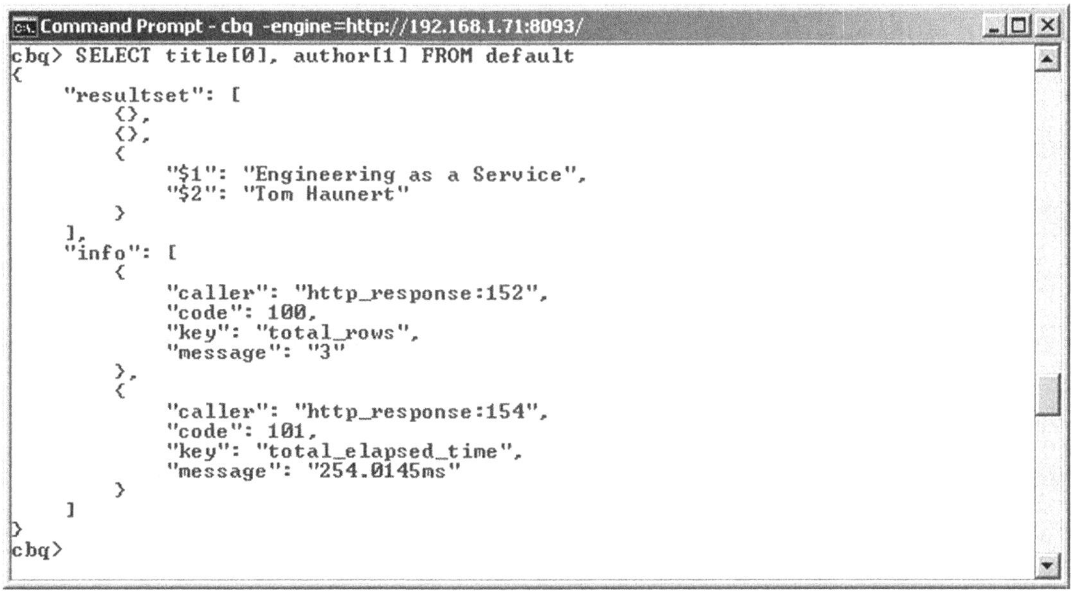

```
Command Prompt - cbq  -engine=http://192.168.1.71:8093/          _ □ ×
cbq> SELECT title[0], author[1] FROM default
{
    "resultset": [
        {},
        {},
        {
            "$1": "Engineering as a Service",
            "$2": "Tom Haunert"
        }
    ],
    "info": [
        {
            "caller": "http_response:152",
            "code": 100,
            "key": "total_rows",
            "message": "3"
        },
        {
            "caller": "http_response:154",
            "code": 101,
            "key": "total_elapsed_time",
            "message": "254.0145ms"
        }
    ]
}
cbq>
```

***Figure 8-14.*** *Selecting Multiple fields*

We have added four different documents to the default bucket: catalog, catalog2, catalog3, and catalog4. If a query to select title is run on the default bucket, all the title fields are returned. Run the following query to select title from all documents.

```
SELECT title FROM default
```

The result set has title fields from all the documents in the default bucket. In two of the documents, catalog & catalog2, the title field has a string value. In the catalog3 document the title field is a nested array, and in the catalog4 document the title field is a nested JSON object as shown in Figure 8-15.

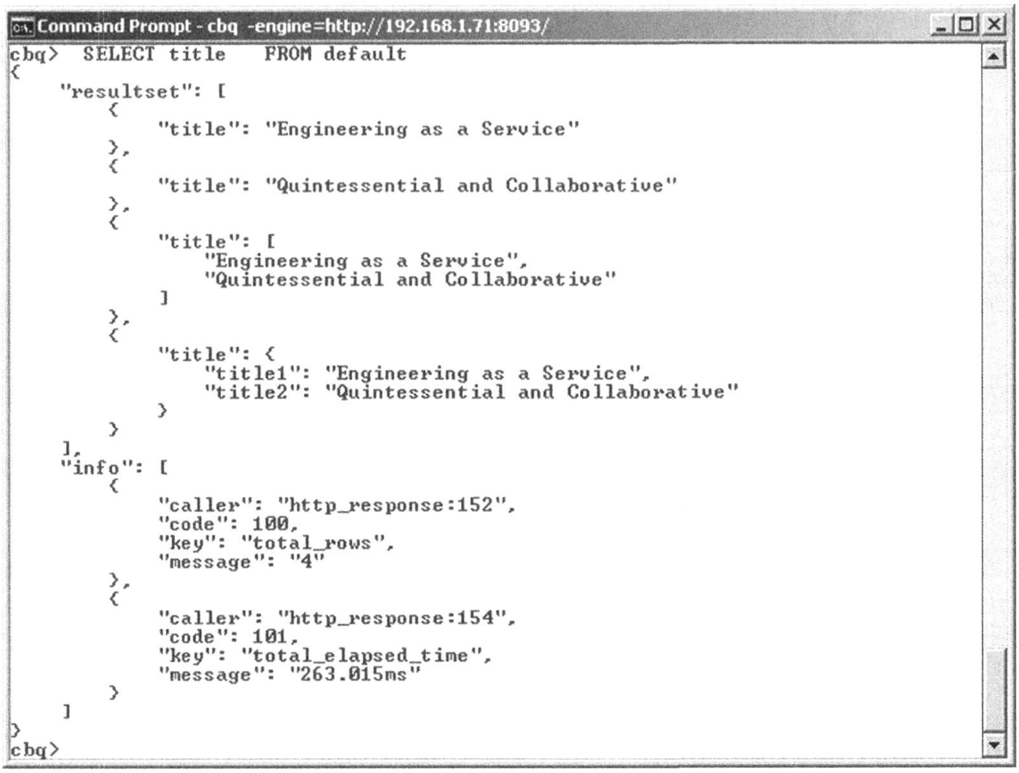

```
Command Prompt - cbq  -engine=http://192.168.1.71:8093/                          _ □ ×
cbq>  SELECT title    FROM default
{
    "resultset": [
        {
            "title": "Engineering as a Service"
        },
        {
            "title": "Quintessential and Collaborative"
        },
        {
            "title": [
                "Engineering as a Service",
                "Quintessential and Collaborative"
            ]
        },
        {
            "title": {
                "title1": "Engineering as a Service",
                "title2": "Quintessential and Collaborative"
            }
        }
    ],
    "info": [
        {
            "caller": "http_response:152",
            "code": 100,
            "key": "total_rows",
            "message": "4"
        },
        {
            "caller": "http_response:154",
            "code": 101,
            "key": "total_elapsed_time",
            "message": "263.015ms"
        }
    ]
}
cbq>
```

***Figure 8-15.*** *Selecting 'title' fields from all Documents from the default Bucket*

The query to select title fields may be filtered using a WHERE clause. For example, the following query selects only those documents in which the edition field has value as 'November-December 2013.'

```
SELECT title FROM default WHERE edition = 'November-December 2013'
```

The result set has only those title fields that have the edition field set to 'November-December 2013.'

```
Command Prompt - cbq  -engine=http://192.168.1.71:8093/                          _ □ ×
cbq> SELECT title FROM default WHERE edition = 'November-December 2013'
{
    "resultset": [
        {
            "title": "Quintessential and Collaborative"
        },
        {
            "title": [
                "Engineering as a Service",
                "Quintessential and Collaborative"
            ]
        },
        {
            "title": {
                "title1": "Engineering as a Service",
                "title2": "Quintessential and Collaborative"
            }
        }
    ],
```

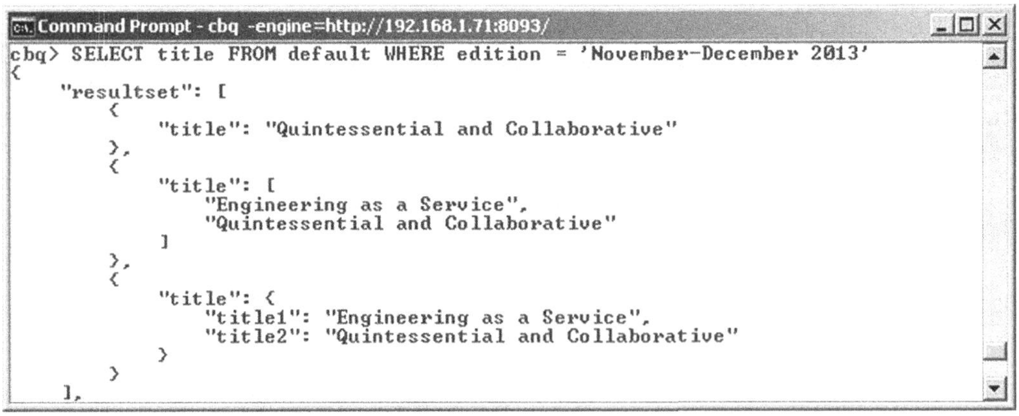

***Figure 8-16.*** *Using the WHERE clause to filter a SELECT query*

# JSON with Nested Objects and Arrays

As the JSON documents stored in Couchbase Server are not based on a fixed-format schema complex JSON documents with nested JSON objects and nested arrays can be queried. Add a document catalog5, which has as value an array in which each array element is a JSON document object.

```
{
  "catalog": [
    {
      "journal": "Oracle Magazine",
      "publisher": "Oracle Publishing",
      "edition": "November-December 2013",
      "title": "Quintessential and Collaborative",
      "author": "Tom Haunert"
    },
    {
      "journal": "Oracle Magazine",
      "publisher": "Oracle Publishing",
      "edition": "November December 2013",
      "title": "Engineering as a Service",
      "author": "David A. Kelly"
    }
  ]
}
```

The document ID catalog5 is shown in Couchbase Console as shown in Figure 8-17.

***Figure 8-17.*** *A JSON document with array elements as JSON documents*

A mixture of . notation and [] notation may be used to select nested arrays and JSON object fields. For example, the following query selects the `title` field in the first array element from the `catalog` field value.

```
SELECT catalog[0].title AS title  FROM default
```

If the document we added is referred to, the first array element has the `title` field as "Quintessential and Collaborative," which is returned in the result set as shown in Figure 8-18.

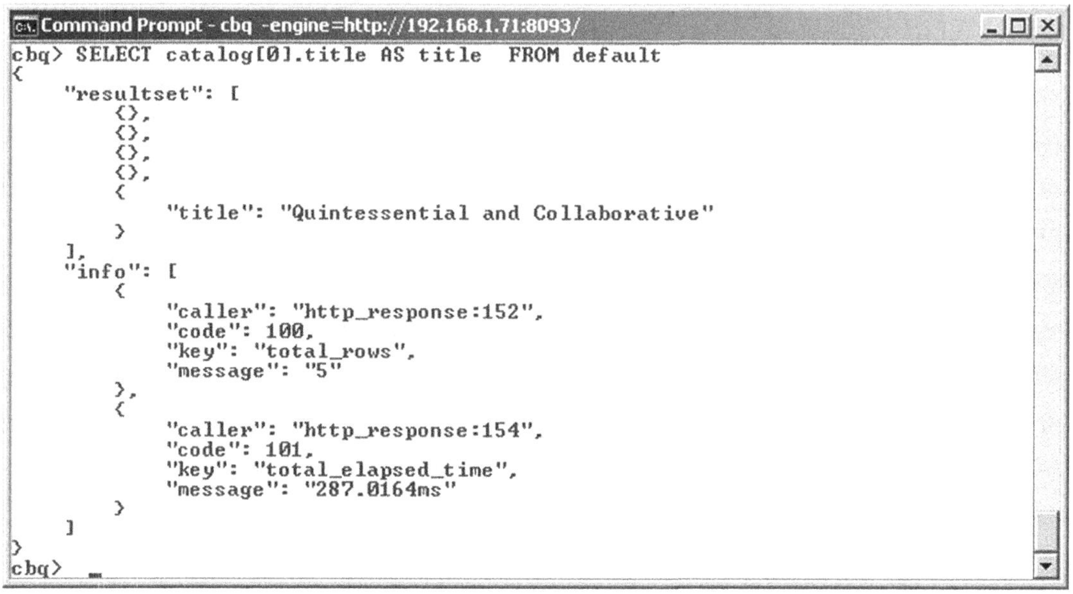
Command Prompt - cbq -engine=http://192.168.1.71:8093/
cbq> SELECT catalog[0].title AS title  FROM default
{
"resultset": [
{},
{},
{},
{},
{
"title": "Quintessential and Collaborative"
}
],
"info": [
{
"caller": "http_response:152",
"code": 100,
"key": "total_rows",
"message": "5"
},
{
"caller": "http_response:154",
"code": 101,
"key": "total_elapsed_time",
"message": "287.0164ms"
}
]
}
cbq>

***Figure 8-18.** Using a mixture of [] and '.' notation*

As another example select the `author` field from the second element in the catalog array.

```
SELECT catalog[1].author AS author  FROM default
```

The `catalog5` document has the `author` field in the second element of the array as "David A. Kelly," which is the value returned in the result set as shown in Figure 8-19.

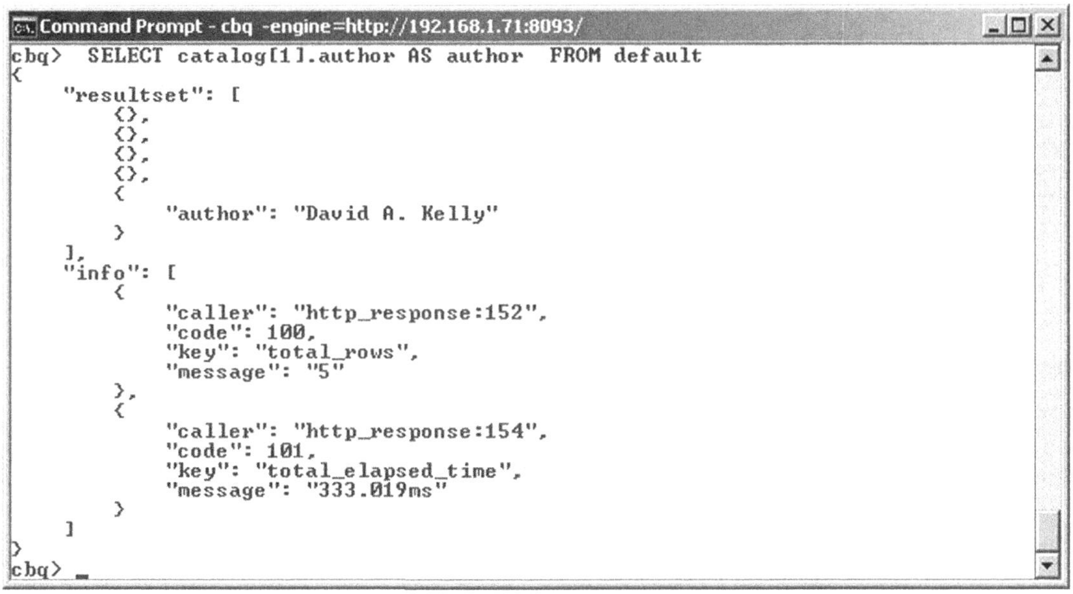

```
Command Prompt - cbq  -engine=http://192.168.1.71:8093/                    _ □ ×
cbq>   SELECT catalog[1].author AS author   FROM default
{
    "resultset": [
        {},
        {},
        {},
        {},
        {
            "author": "David A. Kelly"
        }
    ],
    "info": [
        {
            "caller": "http_response:152",
            "code": 100,
            "key": "total_rows",
            "message": "5"
        },
        {
            "caller": "http_response:154",
            "code": 101,
            "key": "total_elapsed_time",
            "message": "333.019ms"
        }
    ]
}
cbq> _
```

***Figure 8-19.*** *Selecting the author field from the second element in the catalog array*

The JSON document could be even more complex than the preceding example. Add a document catalog6 in which the value of the catalog field is an array in which each element is a JSON object and the journal field of the nested JSON object is an array in which each field is also a JSON object.

```
{
"catalog": [
    {
      "edition": "November-December 2013",
      "journal": [
        { "title": "Engineering as a Service", "author": "David A. Kelly" },
        { "title": "Quintessential and Collaborative", "author": "Tom Haunert"}
      ]
    },
    {
      "edition": "September-October 2013",
      "journal": [
        { "title": "Plug into the Cloud", "author": "David Baum" },
        { "title": "Deploy and Manage Database Clouds", "author": "David Baum" }
      ]
    },
```

```
{
    "edition": "July-August 2013",
    "journal": [
        { "title": "Grow up, Branch Out", "author": "David A. Kelly" },
        { "title": "The CX Factor", "author": "Bob Rhubart" }
    ]
    }
]
}
```

The document with ID `catalog6` is shown in the Couchbase Console as shown in Figure 8-20.

***Figure 8-20.*** *A complex JSON document with nested arrays and objects*

With multiple levels of nesting the path used in the `SELECT` query also has multiple levels of references. As an example, the following query selects the `title` field in the first array element in the `journal` field in the first element in the `catalog` array.

`SELECT catalog[0].journal[0].title  FROM default.`

The first array element in the catalog array is the JSON object with `edition` field as "November-December 2013." The journal field in the array element is also an array in which the first element has the `title` field set to "Engineering as a Service," which is the value returned by the query as shown in Figure 8-21.

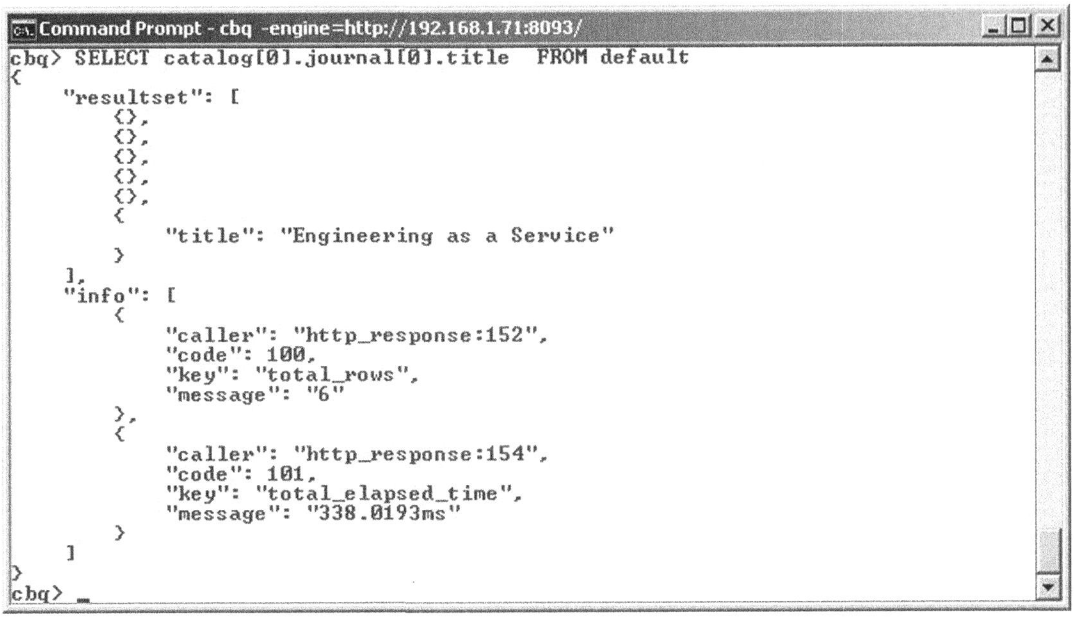

```
Command Prompt - cbq  -engine=http://192.168.1.71:8093/
cbq> SELECT catalog[0].journal[0].title   FROM default
{
    "resultset": [
        {},
        {},
        {},
        {},
        {},
        {
            "title": "Engineering as a Service"
        }
    ],
    "info": [
        {
            "caller": "http_response:152",
            "code": 100,
            "key": "total_rows",
            "message": "6"
        },
        {
            "caller": "http_response:154",
            "code": 101,
            "key": "total_elapsed_time",
            "message": "338.0193ms"
        }
    ]
}
cbq> _
```

*Figure 8-21.*  *Using multiple levels of element dereferencing*

As another example select the value of the edition field in the first array element in the catalog array with the following query. The query is filtered using a WHERE clause.

```
SELECT catalog[0].edition  AS Edition FROM default WHERE catalog[0].edition=
'November-December 2013'
```

The result set includes a JSON object with two fields as Edition. Two fields are returned because another document, catalog5, also has an edition element with value 'November-December 2013' in the first element in the catalog array. As the query runs on all documents in the default bucket, results from all documents are returned.

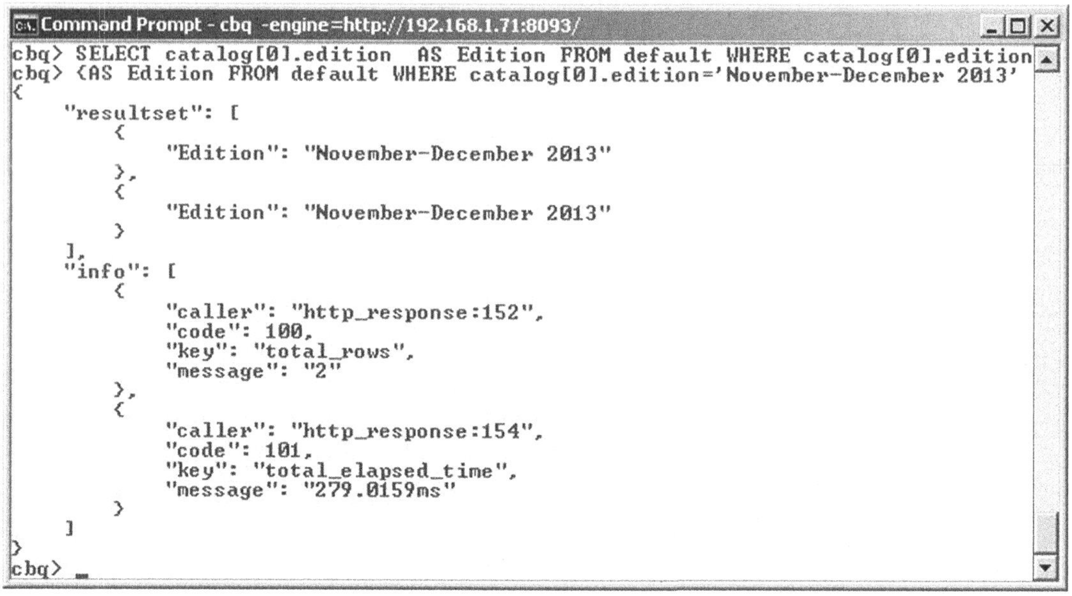

```
Command Prompt - cbq  -engine=http://192.168.1.71:8093/                    _ □ ×
cbq> SELECT catalog[0].edition   AS Edition FROM default WHERE catalog[0].edition
cbq> (AS Edition FROM default WHERE catalog[0].edition='November-December 2013'
{
    "resultset": [
        {
            "Edition": "November-December 2013"
        },
        {
            "Edition": "November-December 2013"
        }
    ],
    "info": [
        {
            "caller": "http_response:152",
            "code": 100,
            "key": "total_rows",
            "message": "2"
        },
        {
            "caller": "http_response:154",
            "code": 101,
            "key": "total_elapsed_time",
            "message": "279.0159ms"
        }
    ]
}
cbq>
```

***Figure 8-22.*** *Including the WHERE clause in array/field dereferencing*

In the preceding queries the result sets had field values as strings. The field values in the JSON resultset could be arrays or JSON objects themselves; a query of a JSON object could return a JSON object. For example, the following query selects the journal field in the first array element from the catalog array using a WHERE clause to include only those documents in which the edition field in the first array element in the catalog array is set to 'November-December 2013'.

```
SELECT catalog[0].journal  FROM default WHERE catalog[0].edition='November-December 2013'
```

The result set includes one of the journal field values as an array with JSON objects as array elements, selected from the catalog6 document. Another journal field is selected from the catalog5 document.

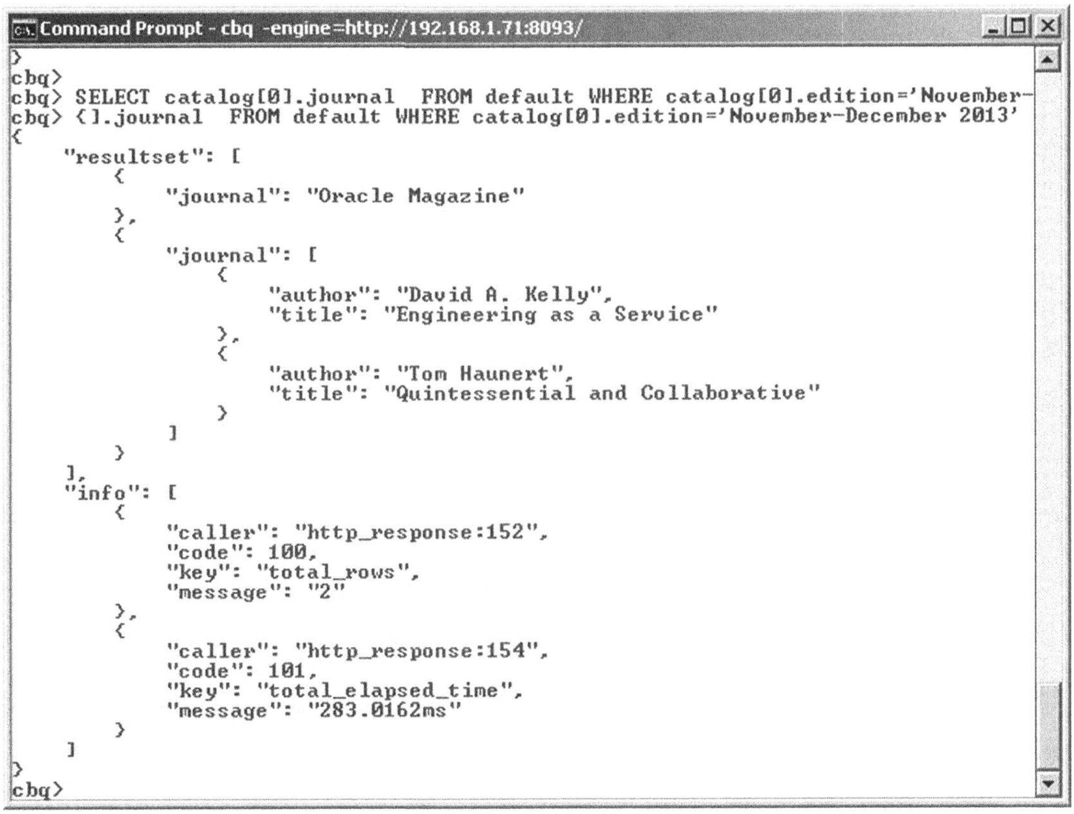

```
cbq>
cbq> SELECT catalog[0].journal  FROM default WHERE catalog[0].edition='November-
cbq> [].journal  FROM default WHERE catalog[0].edition='November-December 2013'
{
    "resultset": [
        {
            "journal": "Oracle Magazine"
        },
        {
            "journal": [
                {
                    "author": "David A. Kelly",
                    "title": "Engineering as a Service"
                },
                {
                    "author": "Tom Haunert",
                    "title": "Quintessential and Collaborative"
                }
            ]
        }
    ],
    "info": [
        {
            "caller": "http_response:152",
            "code": 100,
            "key": "total_rows",
            "message": "2"
        },
        {
            "caller": "http_response:154",
            "code": 101,
            "key": "total_elapsed_time",
            "message": "283.0162ms"
        }
    ]
}
cbq>
```

***Figure 8-23.*** *Resultset with a field value as an array*

All of the standard comparison operators (>, >=, <, <=, =, and !=) are supported in the WHERE clause. Next, we shall run some example queries from the documents in the default bucket. Select the journal field in the first array element in the catalog array in which the title field of the first array element of the journal field is 'Engineering as a Service' as shown in Figure 8-24.

```
SELECT catalog[0].journal  FROM default WHERE catalog[0].journal[0].title='Engineering as a
Service'
```

```
Command Prompt - cbq  -engine=http://192.168.1.71:8093/
cbq> SELECT catalog[0].journal  FROM default WHERE catalog[0].journal[0].title='
cbq> <FROM default WHERE catalog[0].journal[0].title='Engineering as a Service'
<
    "resultset": [
        {
            "journal": [
                {
                    "author": "David A. Kelly",
                    "title": "Engineering as a Service"
                },
                {
                    "author": "Tom Haunert",
                    "title": "Quintessential and Collaborative"
                }
            ]
        }
    ],
    "info": [
        {
            "caller": "http_response:152",
            "code": 100,
            "key": "total_rows",
            "message": "1"
        },
        {
            "caller": "http_response:154",
            "code": 101,
            "key": "total_elapsed_time",
            "message": "281.016ms"
        }
    ]
}
cbq>
```

***Figure 8-24.*** *Select a journal field*

The value returned is an array with each element as a JSON object. The value is selected from the catalog6 document.

Select the edition field in the first element in the catalog array in which the journal field has the title field in its first array element as 'Engineering as a Service'.

```
SELECT catalog[0].edition  FROM default WHERE catalog[0].journal[0].title='Engineering as a
Service'
```

The query returns only one edition field from the catalog6 document as shown in Figure 8-25.

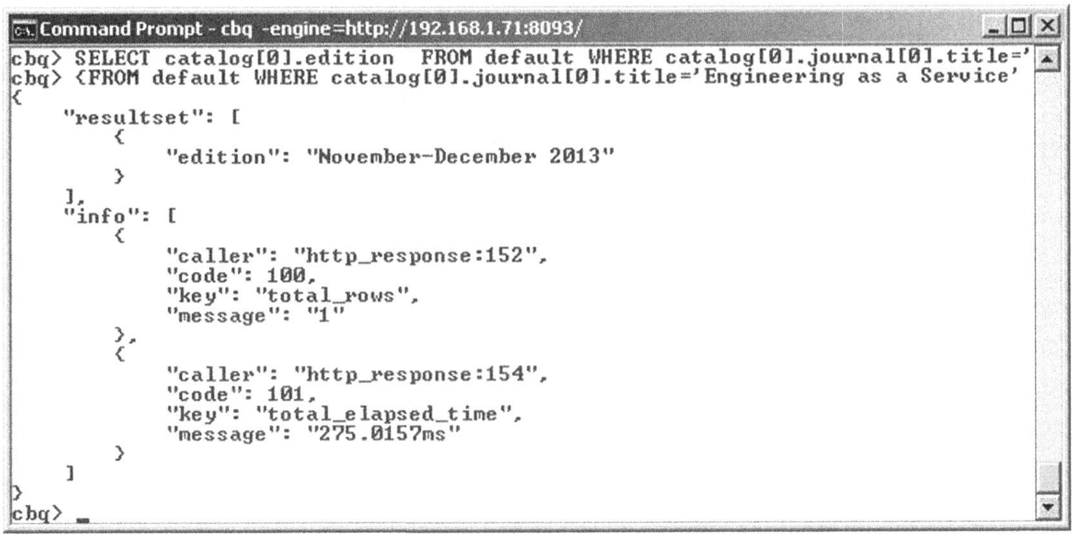

```
Command Prompt - cbq  -engine=http://192.168.1.71:8093/                    _ |□| x|
cbq> SELECT catalog[0].edition  FROM default WHERE catalog[0].journal[0].title='▲
cbq> <FROM default WHERE catalog[0].journal[0].title='Engineering as a Service'
<
    "resultset": [
        {
            "edition": "November-December 2013"
        }
    ],
    "info": [
        {
            "caller": "http_response:152",
            "code": 100,
            "key": "total_rows",
            "message": "1"
        },
        {
            "caller": "http_response:154",
            "code": 101,
            "key": "total_elapsed_time",
            "message": "275.0157ms"
        }
    ]
}
cbq> _                                                                          ▼
```

***Figure 8-25.*** *Running a complex SELECT query*

Select journal, publisher, title, author field values from documents in which the edition field is set to "November-December 2013."

```
SELECT journal, publisher, title, author from default WHERE edition="November-December 2013"
```

The result set array has multiple JSON documents as elements as multiple documents as shown in Figure 8-26.

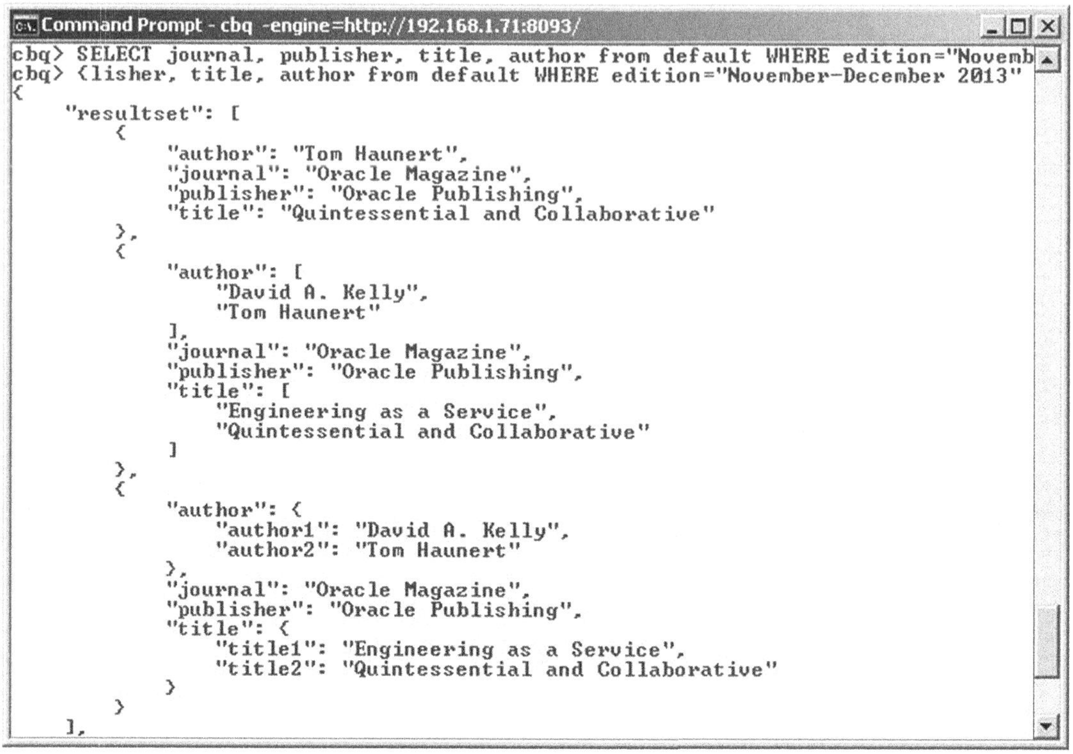

```
Command Prompt - cbq  -engine=http://192.168.1.71:8093/                    _ □ ×
cbq> SELECT journal, publisher, title, author from default WHERE edition="Novemb
cbq> (lisher, title, author from default WHERE edition="November-December 2013"
<
    "resultset": [
        {
            "author": "Tom Haunert",
            "journal": "Oracle Magazine",
            "publisher": "Oracle Publishing",
            "title": "Quintessential and Collaborative"
        },
        {
            "author": [
                "David A. Kelly",
                "Tom Haunert"
            ],
            "journal": "Oracle Magazine",
            "publisher": "Oracle Publishing",
            "title": [
                "Engineering as a Service",
                "Quintessential and Collaborative"
            ]
        },
        {
            "author": {
                "author1": "David A. Kelly",
                "author2": "Tom Haunert"
            },
            "journal": "Oracle Magazine",
            "publisher": "Oracle Publishing",
            "title": {
                "title1": "Engineering as a Service",
                "title2": "Quintessential and Collaborative"
            }
        }
    ],
```

***Figure 8-26.*** *Multiple JSON documents in the resultset*

The WHERE clause is useful when the result set is required to be filtered. One of the documents has an edition set to "November December 2013" instead of "November-December 2013." The journal, publisher, title, and author fields from the document may be selected with the following query.

```
SELECT journal, publisher, title, author from default WHERE edition="November December 2013"
```

The filtered result set includes one row as shown in Figure 8-27.

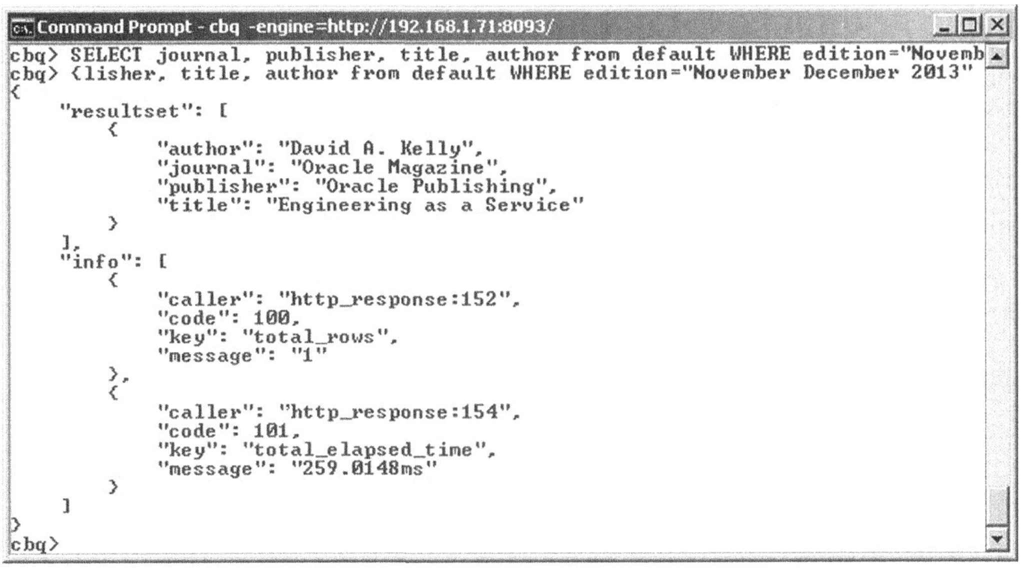

**Figure 8-27.** *Using the WHERE clause in a SELECT query to get one row*

Complex JSON structures may be selected from a document with complex JSON structures using a query with a top-level path. For example, the following query selects the first element in the catalog array from all documents in which the first element in the catalog array has the edition field set to 'November-December 2013.'

```
SELECT catalog[0]  FROM default WHERE catalog[0].edition='November-December 2013'
```

The resultset includes complex JSON structures as shown in Figure 8-28.

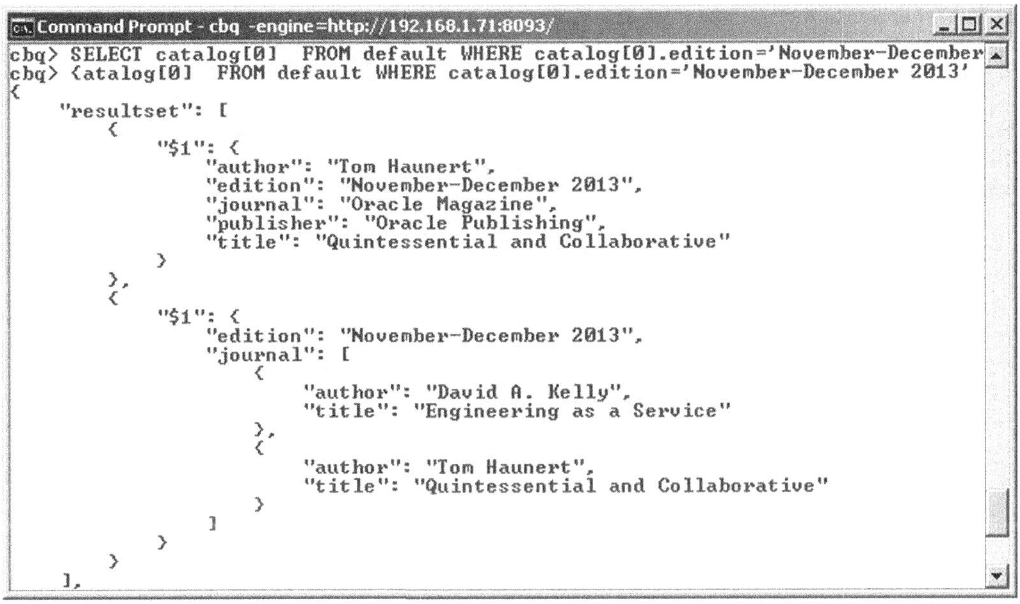

**Figure 8-28.** *Complex JSON structures in the result set*

# Applying Arithmetic & Comparison Operators

N1QL supports the common arithmetic operators +, -, /, * and %. Add a JSON document that has an id field value as an integer on which we may apply the arithmetic operator.

```
{
  "id": 1,
  "journal": "Oracle Magazine"
}
```

The document with id catalog7 is shown in the Couchbase Console in Figure 8-29.

***Figure 8-29.*** *A JSON Document with a numeric id*

The following SELECT query makes use of the * operator on the value of the id field.

```
SELECT id*10, journal from default WHERE journal="Oracle Magazine"
```

The id field value returned is multiplied by 10 in the resultset. The WHERE clause includes all documents in which the journal field is set to "Oracle Magazine" regardless of whether the document has an id field or not as shown in Figure 8-30.

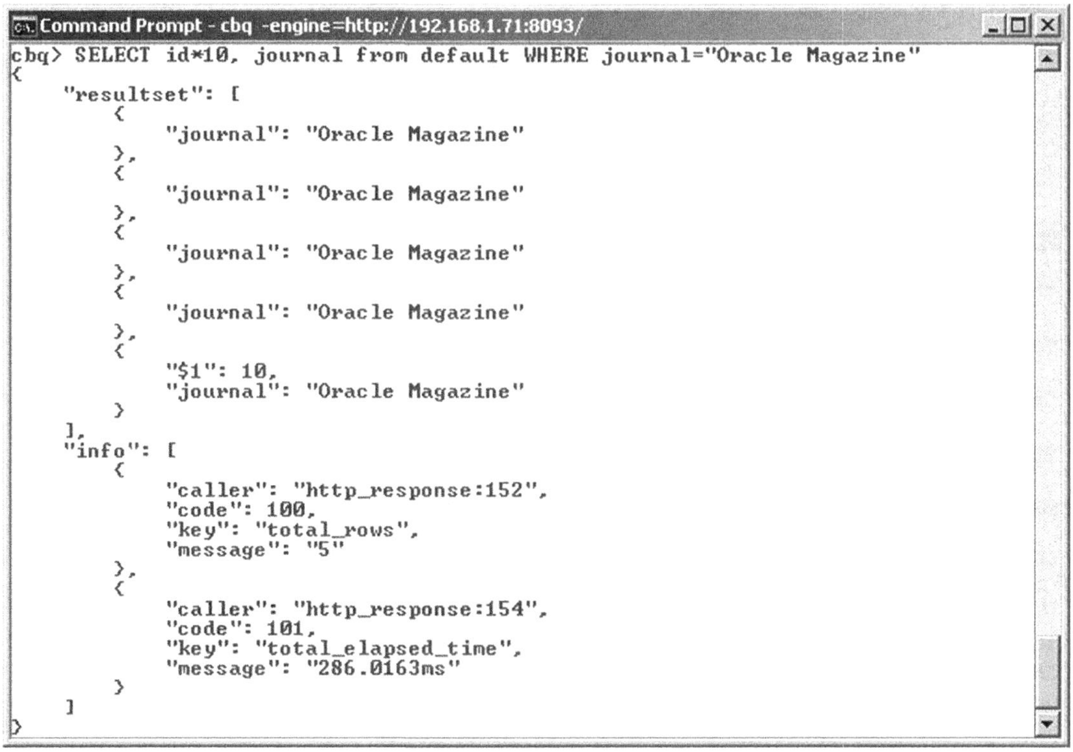

```
Command Prompt - cbq  -engine=http://192.168.1.71:8093/                    _ □ ×
cbq> SELECT id*10, journal from default WHERE journal="Oracle Magazine"
{
    "resultset": [
        {
            "journal": "Oracle Magazine"
        },
        {
            "journal": "Oracle Magazine"
        },
        {
            "journal": "Oracle Magazine"
        },
        {
            "journal": "Oracle Magazine"
        },
        {
            "$1": 10,
            "journal": "Oracle Magazine"
        }
    ],
    "info": [
        {
            "caller": "http_response:152",
            "code": 100,
            "key": "total_rows",
            "message": "5"
        },
        {
            "caller": "http_response:154",
            "code": 101,
            "key": "total_elapsed_time",
            "message": "286.0163ms"
        }
    ]
}
```

*Figure 8-30.* *Using arithmetic Operators*

N1QL also supports the comparison operators >, >=, <, <=, =, and ! =. To demonstrate the use of the comparison operators, add some documents (key-value pairs) with numerical values as listed in Figure 8-31.

| ID | Content |
| --- | --- |
| KV | { "key": 5.5, "value": 11122013 } |
| KV2 | { "key": 4.5, "value": 10 } |
| KV3 | { "key": 10, "value": 10 } |

*Figure 8-31.* *Key/value pairs with numeric values*

Run the following query to select the value in which key is greater than 5 as denoted by the comparison operator >.

```
SELECT value FROM default        WHERE key > 5
```

The result set includes all the value fields greater than 5 as shown in Figure 8-32.

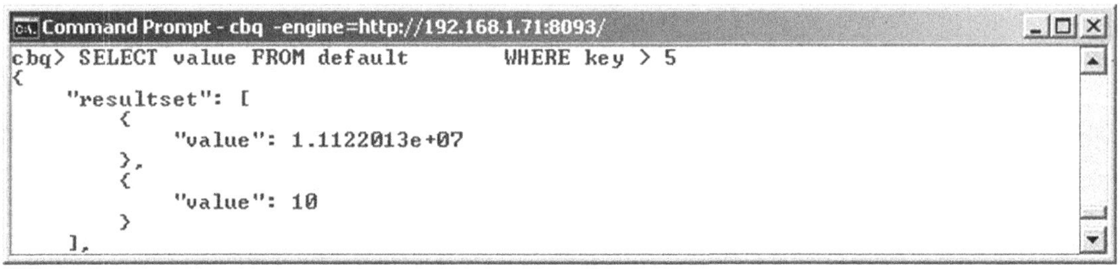

```
Command Prompt - cbq  -engine=http://192.168.1.71:8093/                    _ □ ×
cbq> SELECT value FROM default          WHERE key > 5
{
    "resultset": [
        {
            "value": 1.1122013e+07
        },
        {
            "value": 10
        }
    ],
```

**Figure 8-32.** *Using the comparison Operator >*

# Applying ROUND() and TRUNC() Functions

N1QL supports the built-in functions ROUND() and TRUNC(). The ROUND() function rounds off a value and the TRUNC() function truncates a value. Run the following query to select rounded-off key and truncated value from all the documents.

```
SELECT ROUND(key), TRUNC(value/7) from default
```

Three of the JSON objects returned by the query include the numerical key-value pairs in which the key is rounded off and the value is truncated as shown in Figure 8-33. The other documents also return key-value pairs with both the key and the value fields set to null.

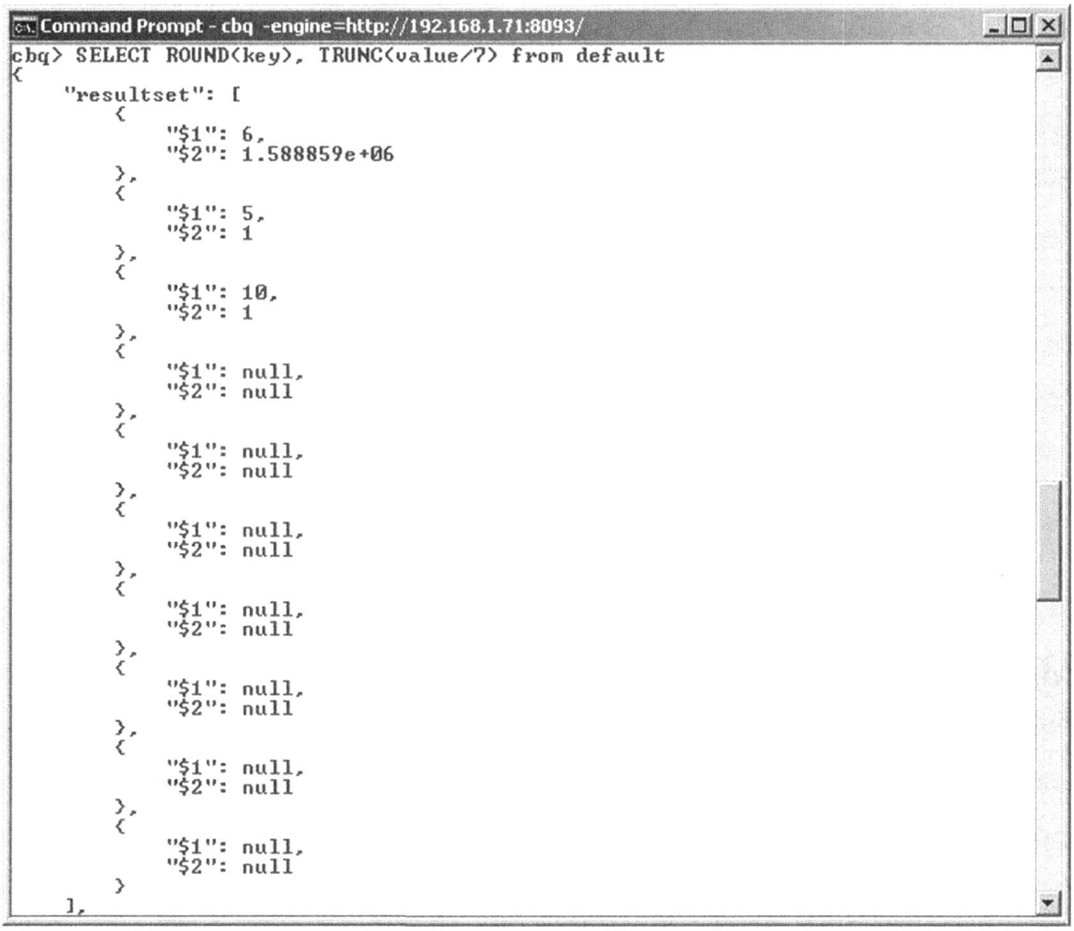

```
Command Prompt - cbq  -engine=http://192.168.1.71:8093/              _ □ ×
cbq> SELECT ROUND(key), TRUNC(value/7) from default
{
    "resultset": [
        {
            "$1": 6,
            "$2": 1.588859e+06
        },
        {
            "$1": 5,
            "$2": 1
        },
        {
            "$1": 10,
            "$2": 1
        },
        {
            "$1": null,
            "$2": null
        },
        {
            "$1": null,
            "$2": null
        },
        {
            "$1": null,
            "$2": null
        },
        {
            "$1": null,
            "$2": null
        },
        {
            "$1": null,
            "$2": null
        },
        {
            "$1": null,
            "$2": null
        },
        {
            "$1": null,
            "$2": null
        }
    ],
```

*Figure 8-33.*  *Using ROUND( ) and TRUC( ) function*

# Concatenating Strings

The || operator can be used to concatenate (link) strings. The following query concatenates the string values returned for the journal, publisher, edition, and title and author fields from documents with edition field set to 'November December 2013'.

```
SELECT journal || " " || publisher || " " || edition || " " || title || " " || author FROM
default WHERE edition='November December 2013'
```

The resultset includes a string value created from joining the string values of individual fields as shown in Figure 8-34.

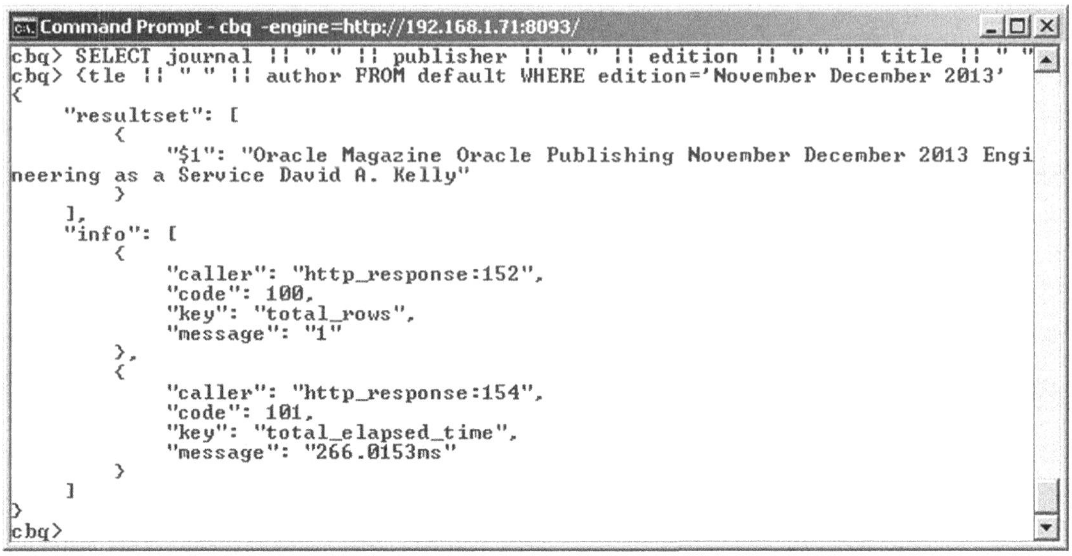

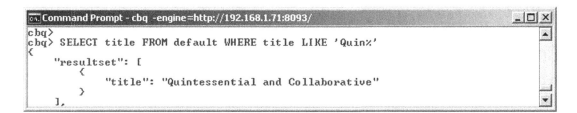

**Figure 8-34.** *Using String concatenation*

# Matching Patterns with LIKE & NOT LIKE

N1QL supports using string patterns in the SELECT statement. The LIKE clause is used to match a pattern. The % can be used as a wildcard to match 0 or more characters. The underscore _ can be used as a wildcard to match exactly one character. The following query uses a string pattern to match titles which start with "Quin."

```
SELECT title FROM default WHERE title LIKE 'Quin%'
```

The title that starts with 'Quin' is returned in the result set as shown in Figure 8-35.

```
Command Prompt - cbq  -engine=http://192.168.1.71:8093/
cbq>
cbq> SELECT title FROM default WHERE title LIKE 'Quin%'
{
    "resultset": [
        {
            "title": "Quintessential and Collaborative"
        }
    ],
```

**Figure 8-35.** *Using the LIKE clause for pattern matching*

The following query selects title field in documents with edition field specified by a string pattern in which the _ character is used.

```
SELECT title FROM default WHERE edition LIKE 'No_ember-_ecember 201_'
```

The resultset includes title fields with edition field value as 'November-December 2013' as shown in Figure 8-36.

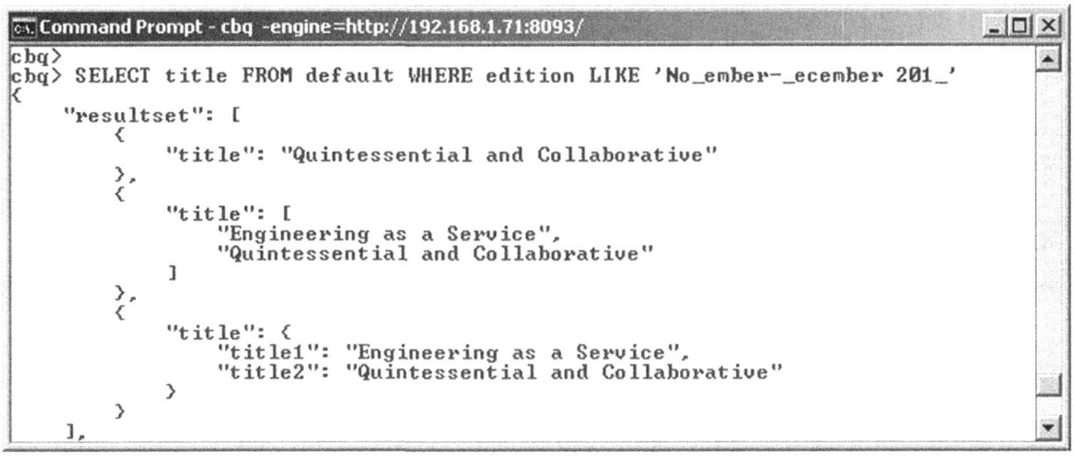

```
Command Prompt - cbq  -engine=http://192.168.1.71:8093/                    _ □ ×
cbq>
cbq> SELECT title FROM default WHERE edition LIKE 'No_ember-_ecember 201_'
{
    "resultset": [
        {
            "title": "Quintessential and Collaborative"
        },
        {
            "title": [
                "Engineering as a Service",
                "Quintessential and Collaborative"
            ]
        },
        {
            "title": {
                "title1": "Engineering as a Service",
                "title2": "Quintessential and Collaborative"
            }
        }
    ],
```

**Figure 8-36.** *Using the _ character in pattern matching*

The NOT LIKE clause can be specified to select fields in which the specified pattern is not matched. In the following query the title field is selected from documents in which the title does not match the pattern Quin%.

```
SELECT title FROM default WHERE title NOT LIKE 'Quin%'
```

The result set includes the title that does not match the pattern as shown in Figure 8-37.

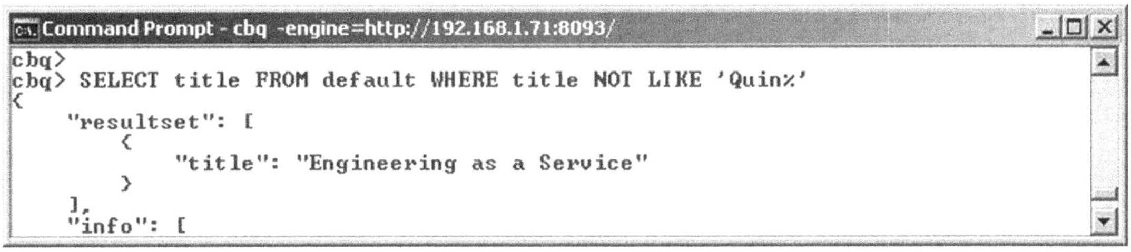

```
Command Prompt - cbq  -engine=http://192.168.1.71:8093/                    _ □ ×
cbq>
cbq> SELECT title FROM default WHERE title NOT LIKE 'Quin%'
{
    "resultset": [
        {
            "title": "Engineering as a Service"
        }
    ],
    "info": [
```

**Figure 8-37.** *Using the NOT LIKE clause*

# Including and Excluding Null and Missing Fields

N1QL supports testing null and missing field values using the IS NULL and IS MISSING clauses. To test null and missing field conditions, add a document KV with a field value as null as shown in Figure 8-38.

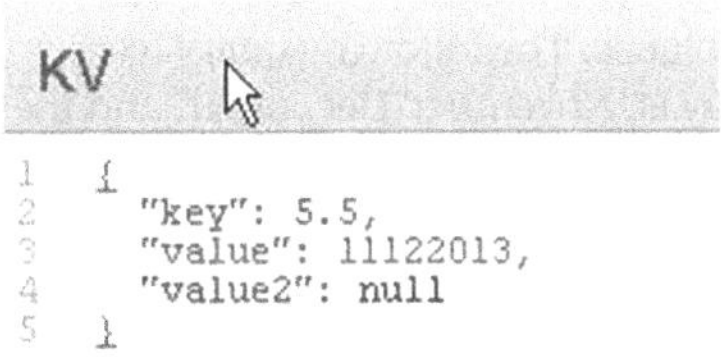

```
KV
1  {
2      "key": 5.5,
3      "value": 11122013,
4      "value2": null
5  }
```

**Figure 8-38.** *JSON Document with a null field value*

Use the following query to select a key from a document in which the value2 field is null.

```
SELECT key from default  WHERE value2 IS null
```

The key-value pairs from the document in which the value2 field is null is returned as shown in Figure 8-39.

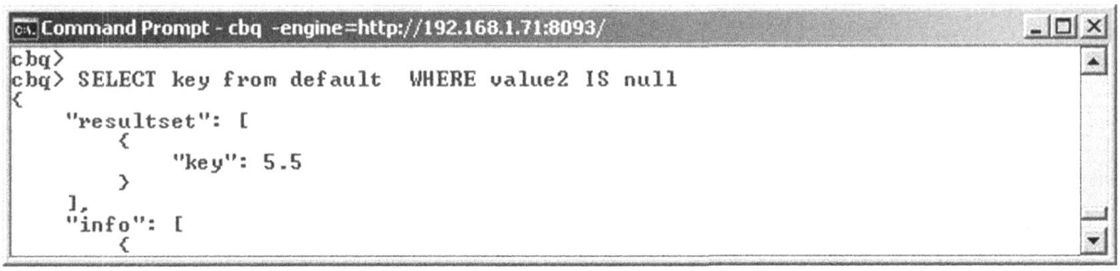

**Figure 8-39.** *Including the null field with the IS NULL clause*

The following query selects key field from documents in which the value2 is missing.

```
SELECT key from default  WHERE value2 IS MISSING
```

The KV2 and KV3 documents are returned in the resultset as shown in Figure 8-40.

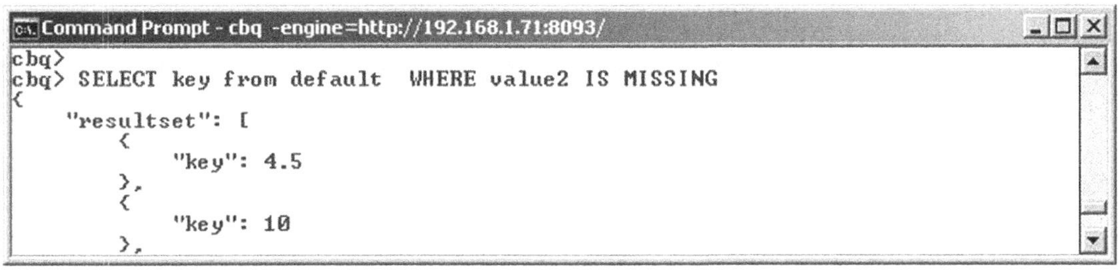

**Figure 8-40.** *Using IS MISSING clauses*

The IS NOT NULL and IS NOT MISSING clauses can be used to test the conditions in which a value is not null or not missing.

# Using Multiple Conditions with AND

Multiple conditions can be tested in selecting documents using the AND clause. The following query tests the condition that the edition field in the first element in the catalog array is 'November-December' and the journal field in the first element in the catalog array is 'Oracle Magazine'.

```
SELECT catalog[0]  FROM default WHERE catalog[0].edition='November-December 2013' AND
catalog[0].journal='Oracle Magazine'
```

The result set contains the catalog[0] array element from the catalog5 document as shown in Figure 8-41.

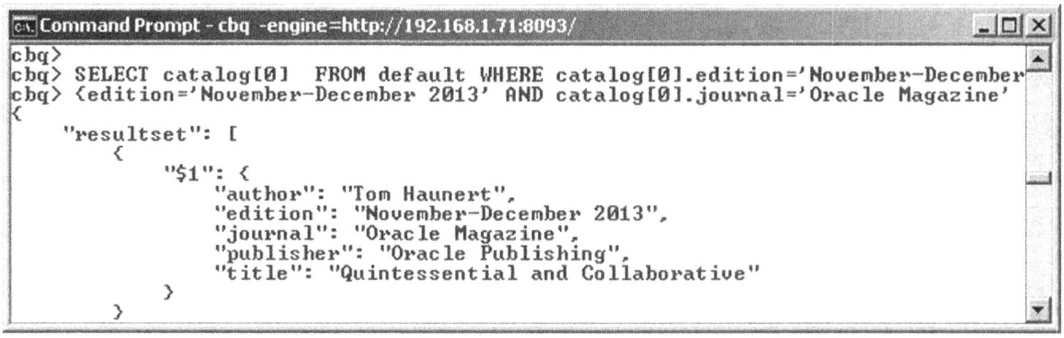

**Figure 8-41.** *Using multiple conditions with the AND clause*

As another example of the AND operator select the catalog[0] array element from documents with the edition in the catalog[0] set to 'November-December' and the catalog[1].edition set to 'September-October 2013.'

```
SELECT catalog[0]  FROM default WHERE catalog[0].edition='November-December 2013' AND
catalog[1].edition='September-October 2013'
```

The first array element from the catalog array in the catalog6 document is returned in the result set as shown in Figure 8-42.

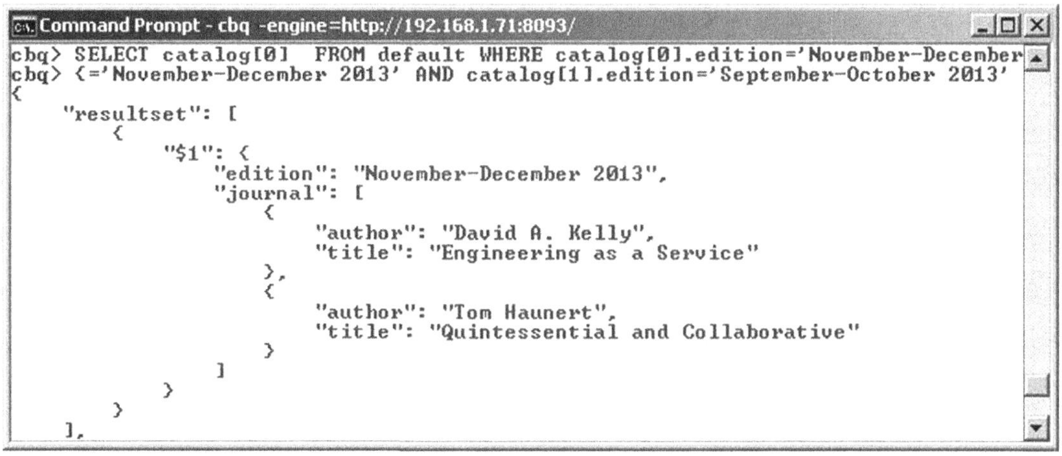

**Figure 8-42.** *Another example of using the AND clause*

# Making Multiple Selections with the OR Clause

The OR clause may be used to select one of the several conditions. The following query selects the title field in which the author is either 'David A. Kelly' or 'Tom Haunert'

```
SELECT title  FROM default WHERE author='David A. Kelly' OR author='Tom Haunert'
```

The result set includes the title field values from the catalog2 and catalog3 documents as shown in Figure 8-43.

```
Command Prompt - cbq  -engine=http://192.168.1.71:8093/                    _ □ ×
cbq> SELECT title   FROM default WHERE author='David A. Kelly' OR author='Tom Hau▲
cbq> <title   FROM default WHERE author='David A. Kelly' OR author='Tom Haunert'
<
     "resultset": [
          {
               "title": "Engineering as a Service"
          },
          {
               "title": "Quintessential and Collaborative"
          }
     ],                                                                         ▼
```

**Figure 8-43.** *Using the OR clause*

# Ordering Result Set

The results can be ordered using the ORDER BY clause to order by a specific field in ascending order by default. Adding the DESC clause orders in descending order instead. The following query selects all "key" and "value" fields and orders them by the "value" field in descending order.

```
SELECT key, value from default   ORDER BY value DESC
```

The result set includes the "key" and "value" fields from the KV, KV2, and KV3 documents ordered by value in descending order as shown in Figure 8-44.

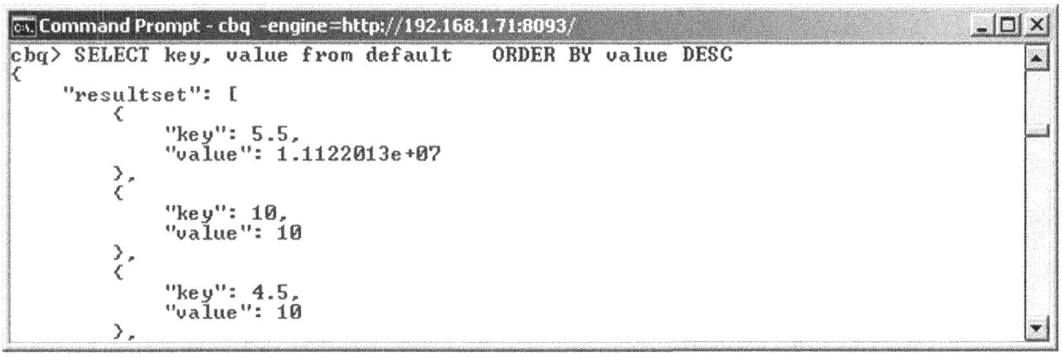

```
Command Prompt - cbq  -engine=http://192.168.1.71:8093/                    _ □ ×
cbq> SELECT key, value from default   ORDER BY value DESC                   ▲
<
     "resultset": [
          {
               "key": 5.5,
               "value": 1.1122013e+07
          },
          {
               "key": 10,
               "value": 10
          },
          {
               "key": 4.5,
               "value": 10
          },                                                                ▼
```

**Figure 8-44.** *Using the ORDER BY clause*

# Using LIMIT and OFFSET to Select a Subset

The number of results returned in the resultset can be limited using the LIMIT clause. When we ran the following query, two results were returned.

```
SELECT catalog[0]  FROM default WHERE catalog[0].edition='November-December 2013'
```

To limit the number of results to one in the preceding query add LIMIT 1 to the query.

```
SELECT catalog[0]  FROM default WHERE catalog[0].edition='November-December 2013' LIMIT 1
```

The result set includes only one result, the first result, instead of two as shown in Figure 8-45.

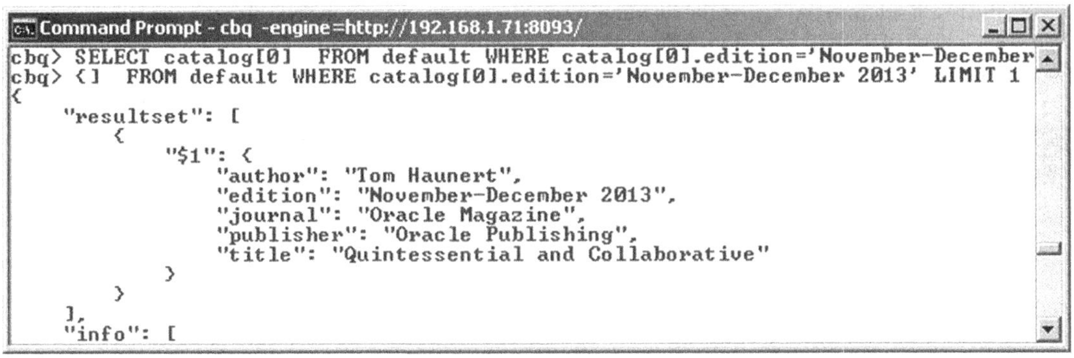

```
Command Prompt - cbq  -engine=http://192.168.1.71:8093/
cbq> SELECT catalog[0]  FROM default WHERE catalog[0].edition='November-December
cbq> <]  FROM default WHERE catalog[0].edition='November-December 2013' LIMIT 1
{
    "resultset": [
        {
            "$1": {
                "author": "Tom Haunert",
                "edition": "November-December 2013",
                "journal": "Oracle Magazine",
                "publisher": "Oracle Publishing",
                "title": "Quintessential and Collaborative"
            }
        }
    ],
    "info": [
```

*Figure 8-45.* *Using the LIMIT clause*

But, if only the second of the two results is required, add the OFFSET clause to the query.

```
SELECT catalog[0]  FROM default WHERE catalog[0].edition='November-December 2013' LIMIT 1
OFFSET 1
```

The resultset includes only the second result as shown in Figure 8-46.

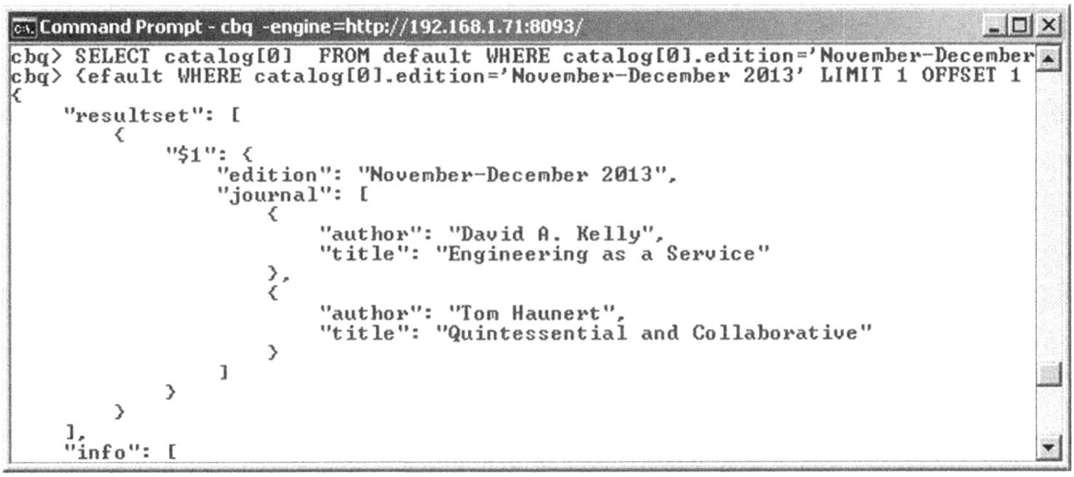

```
Command Prompt - cbq  -engine=http://192.168.1.71:8093/
cbq> SELECT catalog[0]  FROM default WHERE catalog[0].edition='November-December
cbq> <efault WHERE catalog[0].edition='November-December 2013' LIMIT 1 OFFSET 1
{
    "resultset": [
        {
            "$1": {
                "edition": "November-December 2013",
                "journal": [
                    {
                        "author": "David A. Kelly",
                        "title": "Engineering as a Service"
                    },
                    {
                        "author": "Tom Haunert",
                        "title": "Quintessential and Collaborative"
                    }
                ]
            }
        }
    ],
    "info": [
```

*Figure 8-46.* *Using the LIMIT and OFFSET clauses*

# Grouping with GROUP BY

The GROUP BY clause may be used to group multiple results by the value of a common field. The following query returns the catalog field and the document count grouped by the catalog field. The document count is returned as a separate object. Each of the catalog field values is returned as a separate JSON object with the corresponding document count to indicate the number of JSON documents in each group.

```
SELECT catalog, COUNT(*) AS count FROM default GROUP BY catalog
```

In the result set from the query, the document count is returned as a separate object as shown in Figure 8-47. Each of the catalog field values is returned as a separate JSON object with the corresponding document count to indicate the number of JSON documents in each group.

```
Command Prompt - cbq  -engine=http://192.168.1.71:8093/                    _ 8 X
cbq> SELECT catalog, COUNT(*) AS count FROM default GROUP BY catalog      ▲
{
    "resultset": [
        {
            "count": 8
        },
        {
            "catalog": [
                {
                    "author": "Tom Haunert",
                    "edition": "November-December 2013",
                    "journal": "Oracle Magazine",
                    "publisher": "Oracle Publishing",
                    "title": "Quintessential and Collaborative"
                },
                {
                    "author": "David A. Kelly",
                    "edition": "November December 2013",
                    "journal": "Oracle Magazine",
                    "publisher": "Oracle Publishing",
                    "title": "Engineering as a Service"
                }
            ],
            "count": 1
        },
        {
            "catalog": [
                {
                    "edition": "November-December 2013",
                    "journal": [
                        {
                            "author": "David A. Kelly",
                            "title": "Engineering as a Service"
                        },
                        {
                            "author": "Tom Haunert",
                            "title": "Quintessential and Collaborative"
                        }
                    ]
                },
                {
                    "edition": "September-October 2013",
                    "journal": [
                        {
                            "author": "David Baum",
                            "title": "Plug into the Cloud"
                        },
                        {
                            "author": "David Baum",
                            "title": "Deploy and Manage Database Clouds"
                        }
                    ]
                },
                {
                    "edition": "July-August 2013",
                    "journal": [
                        {
```

*Figure 8-47.* *Using the GROUP BY clause*

# Filtering with HAVING

The HAVING clause when added subsequent to the GROUP BY clause filters the resultset using the condition specified in HAVING. The following query adds the HAVING clause to the preceding query.

```
SELECT catalog, COUNT(*) AS count FROM default GROUP BY catalog  HAVING COUNT(*) > 1
```

Only JSON document objects with a count greater than one are returned, which excludes all the catalog field JSON documents as none have a count greater than one as shown in Figure 8-48.

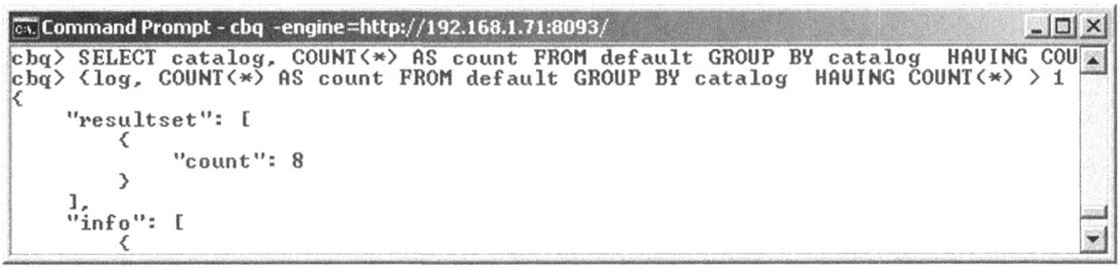

*Figure 8-48.* *Using the HAVING clause*

# Selecting Distinct Values

Duplicate result objects may be returned using the DISTINCT clause. The following query returns distinct edition fields.

```
SELECT DISTINCT edition FROM default
```

Only two distinct fields are returned by the query as shown in Figure 8-49.

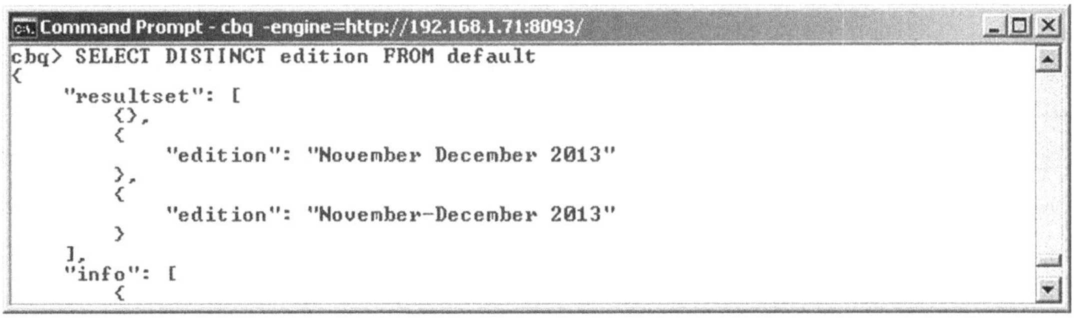

*Figure 8-49.* *Using the DISTINCT clause*

# Summary

In this chapter we discussed the Couchbase Query Language N1QL. We ran the SELECT query to select document/s from Couchbase. We used the WHERE clause in SELECT query to filter the resultset. We also discussed other clauses such as LIKE, NOT LIKE, AND, OR, LIMIT, GROUP BY, and HAVING. We also discussed some commonly used functions such as ROUND() and TRUNC(). In the next chapter we will migrate a MongoDB document to Couchbase.

# CHAPTER 9

■ ■ ■

# Migrating MongoDB

MongoDB is an open source NoSQL database written in C++ with support for dynamic schemas. MongoDB stores documents in JSON-like format called BSON. BSON supports embedding of objects and arrays within other arrays and objects. BSON is lightweight, traversable, and efficient. While MongoDB offers some of the same advantages offered by Couchbase, it also has some limitations. Couchbase has the following advantages over MongoDB.

- *Scalability*. Couchbase is scalable and it is easy to add new servers to a cluster. Couchbase cluster manager is built on top of Erlang/OTP, a proven environment for building fault-tolerant distributed systems. For MongoDB the configuration is fixed; once the shard key, the key to distribute documents between different nodes of a shard cluster, is defined it is fixed and is difficult to change afterwards.

- *Monitoring*. Couchbase comes with a monitoring package while MongoDB requires a subscription. MongoDB can be monitored using the command-line, but the command-line does not provide a graphical user interface.

- *Querying*. Couchbase provides querying using views, which is based on the Map-Reduce concept. A map function and a reduce function may be defined for a query. Couchbase also provides Elasticsearch as a plug-in. MongoDB provides querying based on SQL-like operators with support for indexes and secondary indexes.

- *Management Console*. Couchbase provides a Management GUI Console, which MongoDB doesn't.

In this chapter we shall migrate data from MongoDB to Couchbase Server. The chapter has the following sections.

- Setting Up the Environment
- Creating a Maven Project
- Creating Java Classes
- Configuring the Maven Project
- Creating a BSON Document in MongoDB
- Migrating the MongoDB Document to Couchbase

# Setting Up the Environment

Download the following software for Couchbase Server and MongoDB.

- Couchbase Server Community or Enterprise Edition 3.0.x (or later version) couchbase-server-enterprise_3.0.3-windows_amd64..exe file from http://www.couchbase.com/nosql-databases/downloads. Double-click on the exe file to launch the installer and install Couchbase Server.

- Eclipse IDE for Java EE Developers from http://www.eclipse.org/downloads/.

- MongoDB 3.04 (or a later version) Windows binaries mongodb-win32-x86_64-2008plus-ssl-3.0.4-signed.exe from http://www.mongodb.org/downloads. Double-click on the mongodb-win32-x86_64-2008plus-ssl-3.0.4-signed.exe file to install MongoDB 3.04. Add the bin directory, for example, C:\Program Files\ MongoDB\Server\3.0\bin, to the PATH environment variable.

Start MongoDB server with the following command.

```
>mongod
```

MongoDB server gets started as shown in Figure 9-1.

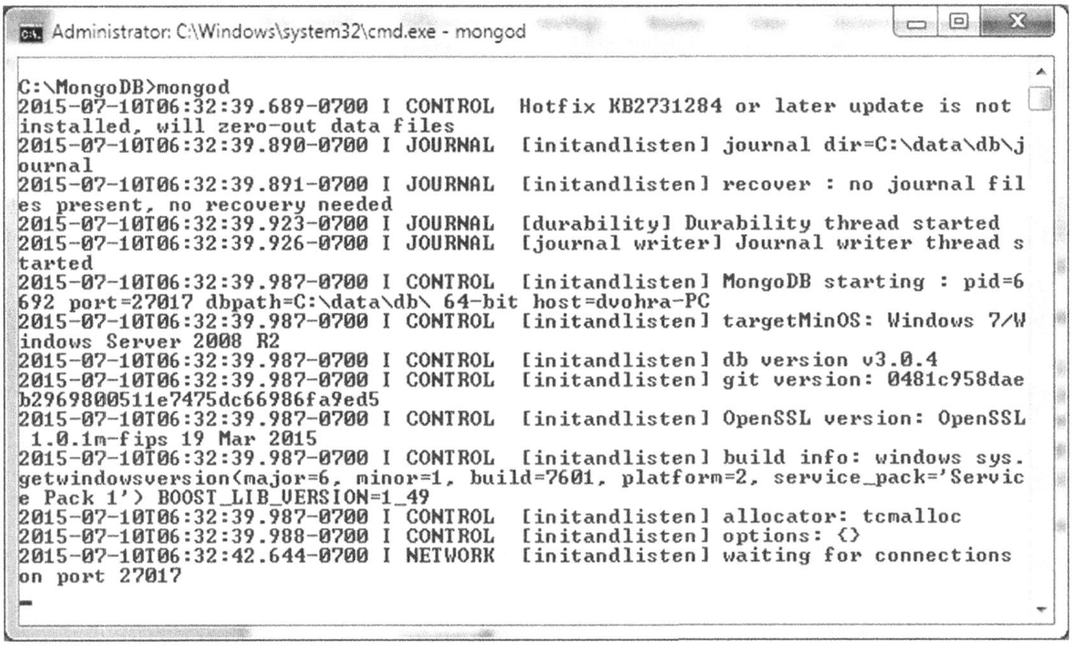

*Figure 9-1. Starting MongoDB*

# Creating a Maven Project

Next, create a Maven project in Eclipse.

1. Select File ➤ New ➤ Other.

2. In the New window, select Maven ➤ Maven Project and click on Next as shown in Figure 9-2.

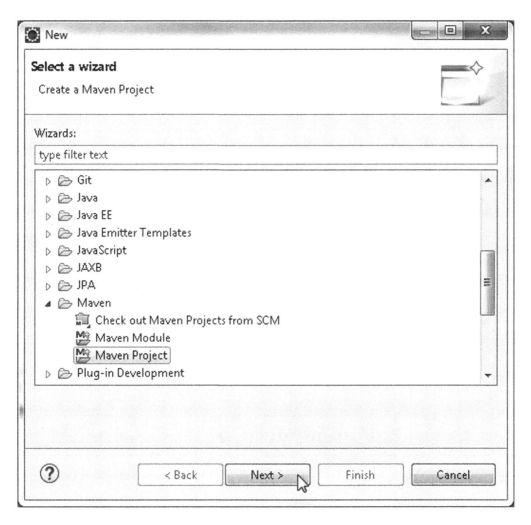

***Figure 9-2.*** *Selecting Maven ➤ Maven Project*

3. In the New Maven Project wizard select the "Create a simple project" check box and the "Use default Workspace location" check box and click on Next as shown in Figure 9-3.

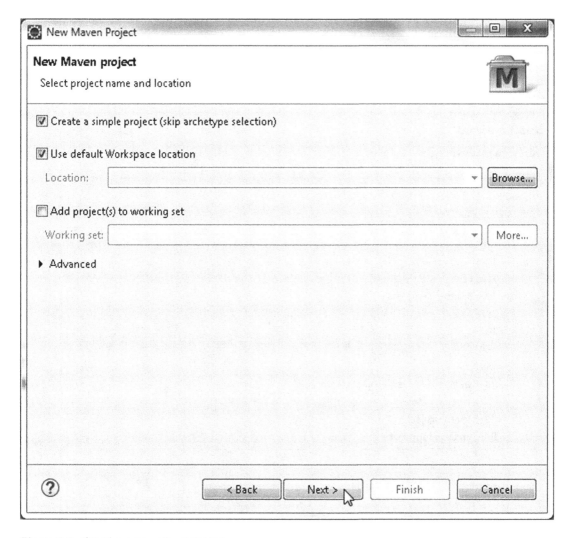

***Figure 9-3.*** *Creating a new Maven Project*

To configure the Maven project, specify the following settings and click Finish as shown in Figure 9-4.

- Group Id: com.couchbase.configuration

- Artifact Id: MongoDBToCouchbase

- Version: 1.0.0

- Packaging: jar

- Name: MongoDBToCouchbase

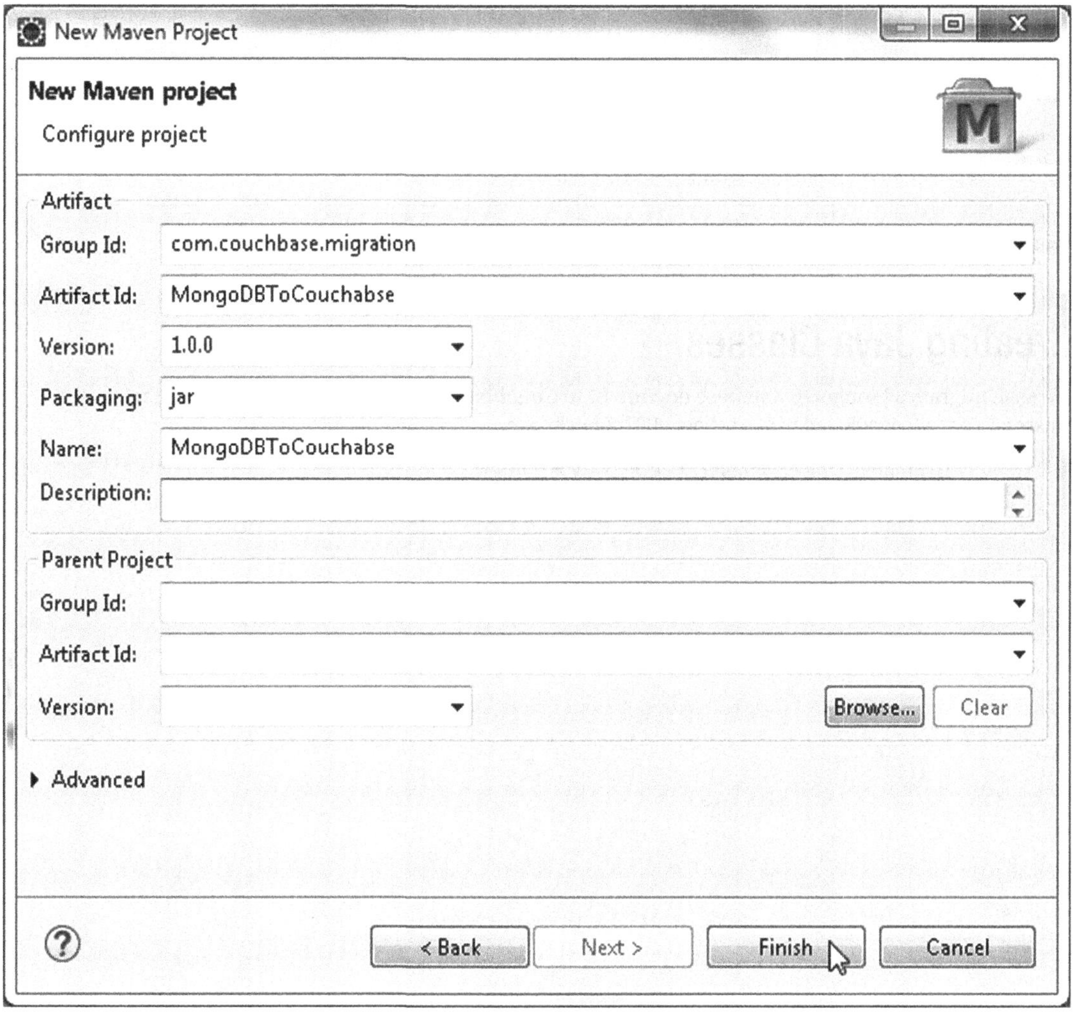

**Figure 9-4.** *Configuring Maven Project*

A Maven project gets added to the Package Explorer in Eclipse as shown in Figure 9-5.

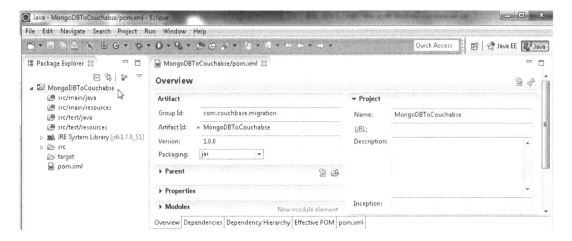

**Figure 9-5.** *Maven Project in Package Explorer*

# Creating Java Classes

We shall migrate a MongoDB database document to Couchbase Server in a Java application. Create two classes: CreateMongoDB and MigrateMongoDBToCouchbase.

1. To create a Java class select File ➤ New ➤ Other.

2. In the New window, select Java ➤ Class and click on Next as shown in Figure 9-6.

***Figure 9-6.*** *Selecting Java ➤ Java Class*

3. In New Java Class wizard select the Source folder and specify Package as couchbase. Specify class Name as CreateMongoDB and click on Finish as shown in Figure 9-7.

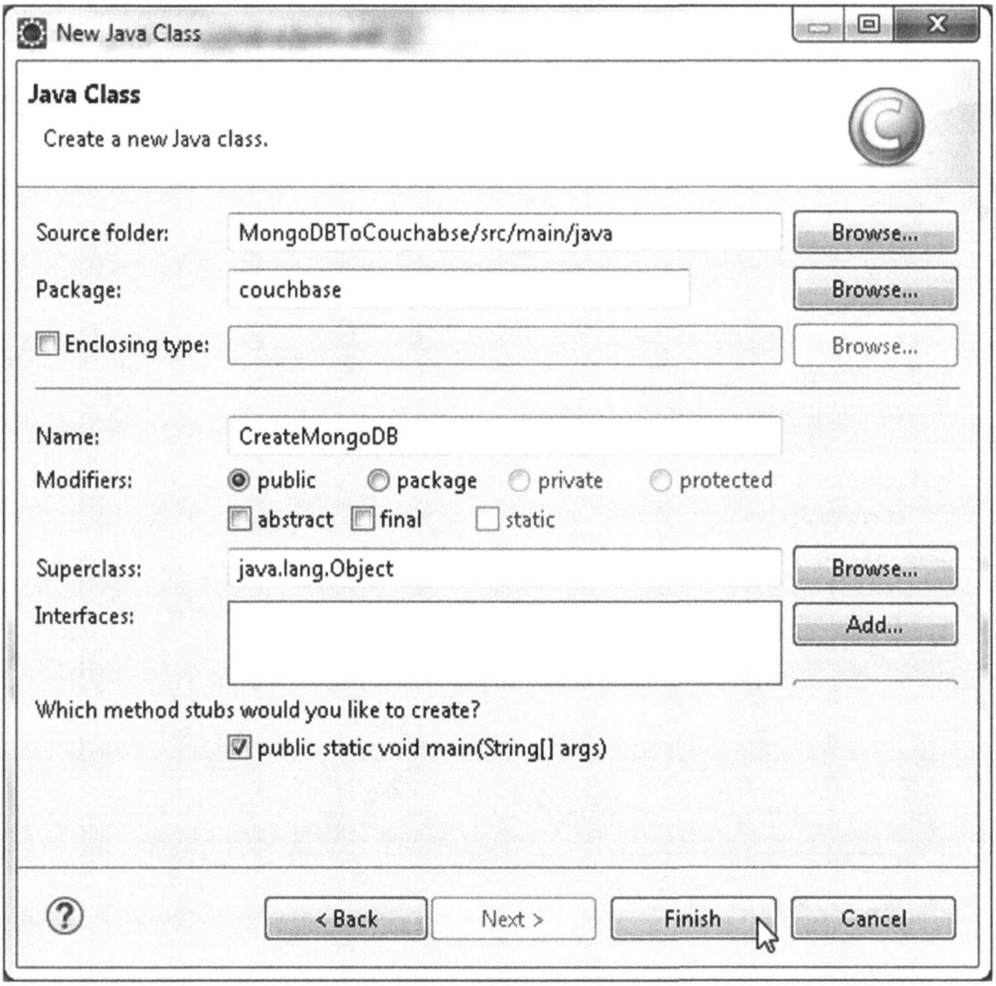

*Figure 9-7.* *New Java Class Wizard*

4. Similarly, add a class MigrateMongoDBToCouchbase as shown in Figure 9-8.

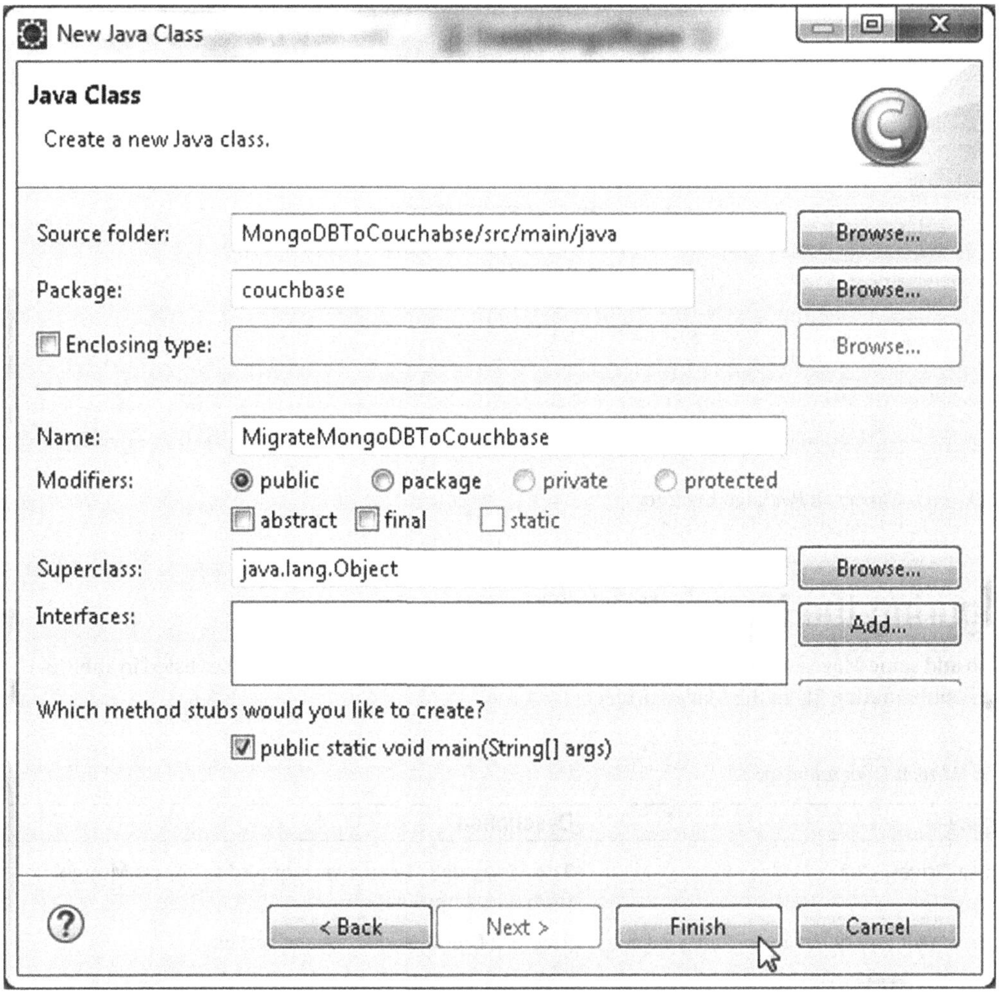

**Figure 9-8.** *Creating MigrateMongoDBToCouchbase Java Class*

The two classes CreateMongoDB and MigrateMongoDBToCouchbase are shown in Package Explorer as shown in Figure 9-9.

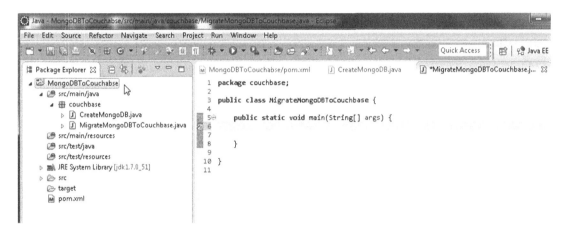

***Figure 9-9.** Java Classes in Package Explorer*

# Configuring the Maven Project

We need to add some Maven dependencies to the project classpath. Add the dependencies listed in Table 9-1 to pom.xml configuration file in the Maven project.

***Table 9-1.** Maven Dependencies*

| Dependency | Description |
| --- | --- |
| Mongo Java Driver 3.0.2 | The MongoDB Java driver required to access MongoDB from a Java application. |
| Couchbase Server Java SDK Client library 2.1.3 | The Java Client to Couchbase Server. |
| Apache Commons BeanUtils 1.9.2 | Utility Jar for Java classes developed with the JavaBeans pattern. |
| Apache Commons Collections 3.2.1 | Java Collections framework provides data structures that accelerate development. |
| Apache Commons Logging 1.2 | An interface for common logging implementations. |
| EZMorph 1.0.6 | Provides conversion from one object to another and is used to convert between non-JSON objects and JSON objects. |

The pom.xml is listed below.

```xml
<project xmlns="http://maven.apache.org/POM/4.0.0"
xmlns:xsi="http://www.w3.org/2001/XMLSchema-instance"
    xsi:schemaLocation="http://maven.apache.org/POM/4.0.0
    http://maven.apache.org/xsd/maven-4.0.0.xsd">
    <modelVersion>4.0.0</modelVersion>
    <groupId>com.couchbase.migration</groupId>
    <artifactId>MongoDBToCouchabse</artifactId>
    <version>1.0.0</version>
    <name>MongoDBToCouchabse</name>
    <dependencies>
        <dependency>
            <groupId>com.couchbase.client</groupId>
            <artifactId>java-client</artifactId>
            <version>2.1.4</version>
        </dependency>
        <dependency>
            <groupId>org.mongodb</groupId>
            <artifactId>mongo-java-driver</artifactId>
            <version>3.0.2</version>
        </dependency>
        <dependency>
            <groupId>commons-beanutils</groupId>
            <artifactId>commons-beanutils</artifactId>
            <version>1.9.2</version>
        </dependency>
        <dependency>
            <groupId>commons-collections</groupId>
            <artifactId>commons-collections</artifactId>
            <version>3.2.1</version>
        </dependency>
        <dependency>
            <groupId>commons-logging</groupId>
            <artifactId>commons-logging</artifactId>
            <version>1.2</version>
        </dependency>
        <dependency>
            <groupId>net.sf.ezmorph</groupId>
            <artifactId>ezmorph</artifactId>
            <version>1.0.6</version>
        </dependency>
    </dependencies>
</project>
```

Select File ➤ Save All to save the pom.xml configuration file. The required jar files get downloaded, and added to the Java build path. To find which Jars have been added to Maven project Java build path, right-click on the project node in Package Explorer and select Properties. In Properties select Java Build Path. The Jars added to the migration project are shown in Figure 9-10.

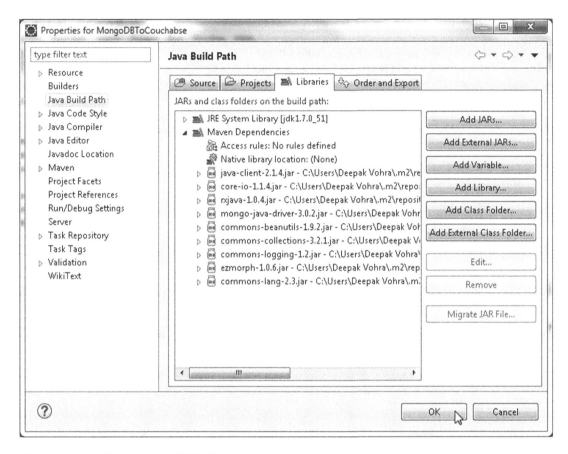

**Figure 9-10.** *Jar Files in Java Build Path*

# Creating a BSON Document in MongoDB

We need to add some data to MongoDB to migrate the data to Couchbase. Next, we shall create a document in MongoDB using the Java application CreateMongoDB. The main packages for MongoDB classes in the MongoDB Java driver are com.mongodb and com.mongodb.client. A MongoDB client to connect to MongoDB server is represented with the com.mongodb.MongoClient class. A MongoClient object provides connection pooling and only one instance is required for the entire instance. The MongoClient class provides several constructors to create an instance from some of which are listed in Table 9-2.

***Table 9-2.*** *MongoClient Class Constructors*

| Constructor | Description |
| --- | --- |
| `MongoClient()` | Creates an instance based on a single MongoDB node for localhost and default port 27017. |
| `MongoClient(String host)` | Creates an instance based on a single MongoDB node with host specified as a host:port String. |
| `MongoClient(String host, int port)` | Creates an instance based on a single MongoDB node using the specified host and port. |
| `MongoClient(List<ServerAddress> seeds)` | Creates an instance using a List of MongoDB servers to select from. The server with the lowest ping time is selected. If the lowest ping time server is down, the next in the list is selected. |
| `MongoClient(List<ServerAddress> seeds, List<MongoCredential> credentialsList)` | Same as the previous version except a List of credentials are provided to authenticate connections to the server/s. |
| `MongoClient(List<ServerAddress> seeds, List<MongoCredential> credentialsList, MongoClientOptions options)` | Same as the previous version except that Mongo client options are also provided. |

Create a `MongoClient` instance using the `MongoClient(List<ServerAddress> seeds)` constructor. Supply "localhost" or the IPv4 address of the host and port as 27017.

```
MongoClient mongoClient = new MongoClient(Arrays.asList(new ServerAddress("192.168.1.71", 27017)));
```

When creating many `MongoClient` instances, all resource usage limits apply per `MongoClient` instance. To dispose of an instance you need to make sure you call `MongoClient.close()` to clean up resources. A logical database in MongoDB is represented with the `com.mongodb.client. MongoDatabase` class. Obtain a `com.mongodb.client. MongoDatabase` instance for the "local" database, which is a default MongoDB database instance, using the `getDatabase(String dbname)` method in `MongoClient` class.

```
MongoDatabase db = mongoClient.getDatabase("local");
```

Some of the Mongo client API has been modified in version 3.0. For example, a database instance is represented with the `MongoDatabase` in 3.0 instead of `com.mongodb.DB` and a database collection in 3.0 is represented with `com.mongodb.client.MongoCollection` instead of `com.mongodb.DBCollection`. MongoDB stores data in collections. Get all collections from the database instance using the `listCollectionNames()` method in `MongoDatabase`.

```
MongoIterable<String> colls = db.listCollectionNames();
```

The `listCollectionNames()` method returns a `MongoIterable<String>` of collections. Iterate over the collection to output the collection names.

```
for (String s : colls) {
System.out.println(s);}
```

Next, create a new DBCollection instance using the getCollection(String collectionName) method in MongoDatabase.

Create a collection of Document instances called catalog.

```
MongoCollection<Document> coll = db.getCollection("catalog");
```

A MongoDB specific BSON object is represented with the org.bson.Document class, which implements the Mapinterface. The Document class provides the following constructors listed in Table 9-3 to create a new instance.

*Table 9-3. Document class Constructors*

| Constructor | Description |
| --- | --- |
| Document() | Creates an empty Document instance. |
| Document(Map<String,Object> map) | Creates a Document instance initialized with a Map. |
| Document(String key, Object value) | Creates a Document instance initialized with a key-value pair. |

The Document class provides some other utility methods, some of which are in Table 9-4.

*Table 9-4. Document class utility Method*

| Method | Description |
| --- | --- |
| append(String key, Object value) | Appends a key-value pair to a Document object and returns a new instance. |
| toString() | Returns a JSON serialization of the object. |

Create a Document instance using the Document(String key, Object value) constructor and use the append(String key, Object val) method to append key-value pairs.

```
Document catalog = new Document("journal", "Oracle Magazine")
.append("publisher", "Oracle Publishing")
.append("edition", "November December 2013")
.append("title", "Engineering as a Service").append("author", "David A. Kelly");
```

The MongoCollection interface provides insertOne(TDocument document, SingleResultCallback<Void> callback) method to add Document/s to a collection. Add the catalog Document to the MongoCollection instance for the catalog collection.

```
coll.insertOne(catalog);
```

The MongoCollection interface also provides overloaded find() method to find a Document instance. Obtain the document added using the find() method. The find() method returns an iterable collection from which we obtain the first document using the first() method.

```
Document dbObj = coll.find().first();
```

Output the Document object found as such and also by iterating over the Set obtained from the Document using the keySet() method. The keySet() method returns a Set<String>. Create an Iterator from the Set<String> using the iterator() method. While the Iterator has elements as determined by the hasNext() method obtain the elements using the next() method. Each element is a key in the Document fetched. Obtain the value for the key using the get(String key) method in Document.

```
System.out.println(dbObj);
Set<String> set = dbObj.keySet();
Iterator iter = set.iterator();
while(iter.hasNext()){
Object obj=    iter.next();
System.out.println(obj);
System.out.println(dbObj.get(obj.toString()));
}
```

Clost the MongoClient instance.

```
mongoClient.close();
```

The CreateMongoDB class is listed below.

```
package couchbase;

import java.util.Arrays;
import java.util.Iterator;
import java.util.Set;
import org.bson.Document;
import com.mongodb.MongoClient;
import com.mongodb.ServerAddress;
import com.mongodb.client.MongoCollection;
import com.mongodb.client.MongoDatabase;
import com.mongodb.client.MongoIterable;

public class CreateMongoDB {

    public static void main(String[] args) {

        MongoClient mongoClient = new MongoClient(
                Arrays.asList(new ServerAddress("localhost", 27017)));
        for (String s : mongoClient.listDatabaseNames()) {
            System.out.println(s);
        }
        MongoDatabase db = mongoClient.getDatabase("local");
        MongoIterable<String> colls = db.listCollectionNames();
        System.out.println("MongoDB Collection Names: ");
        for (String s : colls) {
            System.out.println(s);
        }
        MongoCollection<Document> coll = db.getCollection("catalog");
        Document catalog = new Document("journal", "Oracle Magazine")
                .append("publisher", "Oracle Publishing")
```

247

```
            .append("edition", "November December 2013")
            .append("title", "Engineering as a Service")
            .append("author", "David A. Kelly");
    coll.insertOne(catalog);
    Document dbObj = coll.find().first();
    System.out.println(dbObj);
    Set<String> set = catalog.keySet();
    Iterator<String> iter = set.iterator();
    while (iter.hasNext()) {
        Object obj = iter.next();
        System.out.println(obj);
        System.out.println(dbObj.get(obj.toString()));
    }
    mongoClient.close();

    }

}
```

To run the `CreateMongoDB` application, right-click on the `CreateMongoDB.java` file in Package Explorer and select Run As ➤ Java Application as shown in Figure 9-11.

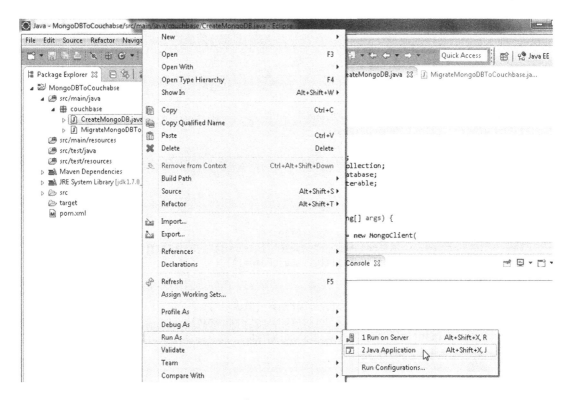

***Figure 9-11.*** *Running CreateMongoDB.java Class*

A new BSON document gets stored in a new collection catalog in MongoDB database. The document stored is also output as such and as key-value pairs as shown in Figure 9-12.

```
<terminated> CreateMongoDB (1) [Java Application] C:\Program Files\Java\jdk1.7.0_51\bin\javaw.exe (Jul 10, 2015, 6:46:15 AM)
INFO: Opened connection [connectionId{localValue:1, serverValue:1}] to localhost:27017
Jul 10, 2015 6:46:20 AM com.mongodb.diagnostics.logging.JULLogger log
INFO: Monitor thread successfully connected to server with description ServerDescription{address=localhost:27017, type=STANDA
LONE, state=CONNECTED, ok=true, version=ServerVersion{versionList=[3, 0, 4]}, minWireVersion=0, maxWireVersion=3, electionId=
null, maxDocumentSize=16777216, roundTripTimeNanos=1032185}
Jul 10, 2015 6:46:20 AM com.mongodb.diagnostics.logging.JULLogger log
INFO: Discovered cluster type of STANDALONE
Jul 10, 2015 6:46:20 AM com.mongodb.diagnostics.logging.JULLogger log
INFO: Opened connection [connectionId{localValue:2, serverValue:2}] to localhost:27017
Loc8r
local
mongo
test
MongoDB Collection Names:
catalog
startup_log
system.indexes
wlslog
Document{{_id=55968c90cd735011a8bfa84a, journal=Oracle Magazine, publisher=Oracle Publishing, edition=November December 2013,
 title=Engineering as a Service, author=David A. Kelly}}
journal
Oracle Magazine
publisher
Oracle Publishing
edition
November December 2013
title
Engineering as a Service
author
David A. Kelly
_id
55968c90cd735011a8bfa84a
Jul 10, 2015 6:46:20 AM com.mongodb.diagnostics.logging.JULLogger log
INFO: Closed connection [connectionId{localValue:2, serverValue:2}] to localhost:27017 because the pool has been closed.
```

***Figure 9-12.*** *Outputting Document Stored in MongoDB*

# Migrating the MongoDB Document to Couchbase

In this section we shall migrate the MongoDB document stored in the previous section to the Couchbase Server. We shall migrate the document in the MigrateMongoDBToCouchbase application. Create a MongoClient instance as discussed in the previous section to add a document.

```
MongoClient mongoClient = new MongoClient(Arrays.asList(new ServerAddress("localhost", 27017)));
```

Create a MongoDatabase object for the local database instance using the getDatabase(String dbname) method in MongoClient. To make a connection to a MongoDB you need to have at the minimum the name of a database to connect to. Using the MongoDatabase instance get the catalog collection as a MongoCollection object. Get a Document instance from the document stored in MongoDB in the previous section using the findOne() method in DBCollection class.

```
MongoDatabase db = mongoClient.getDatabase("local");
MongoCollection<Document> coll = db.getCollection("catalog");
Document catalog = coll.find().first();
```

Add a method called `migrate()` to the `MigrateMongoDBToCouchbase` class and invoke the method in the `main` method. In the `migrate()` method we shall migrate the MongoDB document as represented by the Document object catalog to Couchbase Server. As discussed in Chapter 2 the `CouchbaseCluster` class is used to connect to a Couchbase cluster. Create a `CouchbaseCluster` instance using the static method `create()`.

```
Cluster cluster = CouchbaseCluster.create();
```

A Bucket class instance represents a connection to a data bucket in Couchbase Server. Create a `Bucket` instance using the `openBucket()` method. To connect to the `default` bucket the bucket name may be optionally supplied as an argument to the `openBucket()` method.

```
Bucket defaultBucket = cluster.openBucket("default");
```

Couchbase Server stores documents as JSON. Next, we shall construct JSON strings from data retrieved for MongoDB and store the JSON string in Couchbase. The Document instance catalog is the document fetched from the MongoDB database. The `keySet()` method in Document returns a Set<String> of the BSON keys. Create an `Iterator` from the Set<String> to iterate over the Set using the `iterator()` method.

```
Iterator<String> iter = set.iterator();
```

The `com.couchbase.client.java.document.json.JsonObject` class represents a JSON document in Couchbase Server. We shall use the `JsonObject` class to create a JSON object representation of the Document object and subsequently store the `String` created from `JsonObject` in Couchbase. Create an instance of `JsonObject` using the class method empty().

```
JsonObject catalogObj = JsonObject.empty();
```

While the `Iterator` has elements as determined by the `hasNext()` method obtain the elements using the `next()` method with each element being a key in the Document fetched. Obtain the value for the key using the `get(String key)` method in Document. Use the `put(String key, String value)` method in `JsonObject` to put the column name and column value in the `JsonObject` object.

```
while (iter.hasNext()) {
    String columnName = iter.next().toString();
    String value = catalog.get(columnName.toString()).toString();
    catalogObj.put(columnName, value);
}
```

The Bucket class provides overloaded insert methods to store documents to Couchbase Server. Add the `JsonObject` instance to the `default` bucket using the `Bucket` instance and the `insert(D document)` method. Create an instance of `JsonDocument` to add with the `static` method `JsonDocument.create(java.lang.String id, JsonObject content)`.

```
JsonDocument document = defaultBucket.insert(JsonDocument.create("catalog", catalogObj));
```

The MigrateMongoDBToCouchbase class is listed below.

```
package couchbase;

import java.util.Arrays;
import java.util.Iterator;
import java.util.Set;
import org.bson.Document;
import com.couchbase.client.java.Bucket;
import com.couchbase.client.java.Cluster;
import com.couchbase.client.java.CouchbaseCluster;
import com.couchbase.client.java.document.JsonDocument;
import com.couchbase.client.java.document.json.JsonObject;
import com.mongodb.MongoClient;
import com.mongodb.ServerAddress;
import com.mongodb.client.MongoCollection;
import com.mongodb.client.MongoDatabase;

public class MigrateMongoDBToCouchbase {
    private static Document catalog;

    public static void main(String[] args) {

        MongoClient mongoClient = new MongoClient(
                Arrays.asList(new ServerAddress("localhost", 27017)));
        MongoDatabase db = mongoClient.getDatabase("local");
        MongoCollection<Document> coll = db.getCollection("catalog");
        catalog = coll.find().first();
        migrate();
        mongoClient.close();
    }

    private static void migrate() {
        Cluster cluster = CouchbaseCluster.create();
        Bucket defaultBucket = cluster.openBucket("default");

        Set<String> set = catalog.keySet();
        Iterator<String> iter = set.iterator();
        JsonObject catalogObj = JsonObject.empty();
        while (iter.hasNext()) {
            String columnName = iter.next().toString();
            String value = catalog.get(columnName.toString()).toString();
            catalogObj.put(columnName, value);
        }
        JsonDocument document = defaultBucket.insert(JsonDocument.create(
                "catalog", catalogObj));

        System.out.println("Set Succeeded");
    }

}
```

To run the `MigrateMongoDBToCouchbase` application, right-click on the `MigrateMongoDBToCouchbase` source file in the Package Explorer and select Run As ➤ Java Application as shown in Figure 9-13.

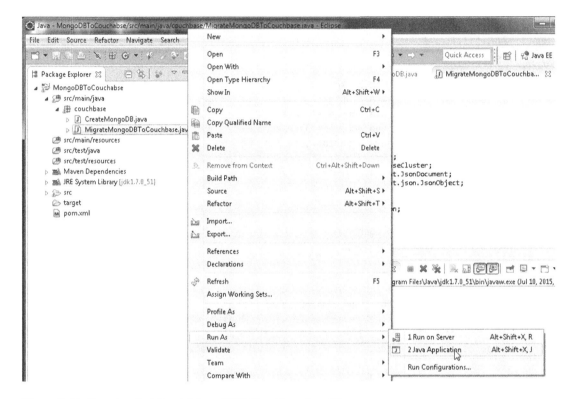

***Figure 9-13.*** *Running the MigrateMongoDBToCouchbase.java Class*

As the output from the application indicates the document fetched from MongoDB gets set in Couchbase Server as shown in Figure 9-14.

```
<terminated> MigrateMongoDBToCouchbase (1) [Java Application] C:\Program Files\Java\jdk1.7.0_51\bin\javaw.exe (Jul 10,
INFO: Opened connection [connectionId{localValue:2, serverValue:4}] to localhost:27017
Jul 10, 2015 7:00:37 AM com.couchbase.client.core.env.DefaultCoreEnvironment <init>
INFO: ioPoolSize is less than 3 (2), setting to: 3
Jul 10, 2015 7:00:37 AM com.couchbase.client.core.env.DefaultCoreEnvironment <init>
INFO: computationPoolSize is less than 3 (2), setting to: 3
Jul 10, 2015 7:00:37 AM com.couchbase.client.core.CouchbaseCore <init>
INFO: CouchbaseEnvironment: {sslEnabled=false, sslKeystoreFile='null', sslKeystorePassword='
ort=8093, bootstrapHttpEnabled=true, bootstrapCarrierEnabled=true, bootstrapHttpDirectPort=8
bootstrapCarrierDirectPort=11210, bootstrapCarrierSslPort=11207, ioPoolSize=3, computationPo
4, requestBufferSize=16384, kvServiceEndpoints=1, viewServiceEndpoints=1, queryServiceEndpoi
coreScheduler=CoreScheduler, eventBus=DefaultEventBus, packageNameAndVersion=couchbase-java-
bled=false, retryStrategy=BestEffort, maxRequestLifetime=75000, retryDelay=ExponentialDelay{
0, upper=100000}, reconnectDelay=ExponentialDelay{growBy 1.0 MILLISECONDS; lower=32, upper=4
ntialDelay{growBy 1.0 MICROSECONDS; lower=10, upper=100000}, keepAliveInterval=30000, autore
bled=true, queryTimeout=75000, viewTimeout=75000, kvTimeout=2500, connectTimeout=5000, disco
=false}
Jul 10, 2015 7:00:39 AM com.couchbase.client.core.node.CouchbaseNode$1 call
INFO: Connected to Node 127.0.0.1
Jul 10, 2015 7:00:39 AM com.couchbase.client.core.config.DefaultConfigurationProvider$6 call
INFO: Opened bucket default
Set Succeeded
Jul 10, 2015 7:00:39 AM com.mongodb.diagnostics.logging.JULLogger log
INFO: Closed connection [connectionId{localValue:2, serverValue:4}] to localhost:27017 becau
```

*Figure 9-14.* *Output from running MigrateMongoDBToCouchbase.java Class*

Log in to the Administration Console for Couchbase Server and click on the Data Buckets node. The default bucket should be listed and the catalog ID document should be listed in the default bucket. Click on Edit Document to display the document as shown in Figure 9-15.

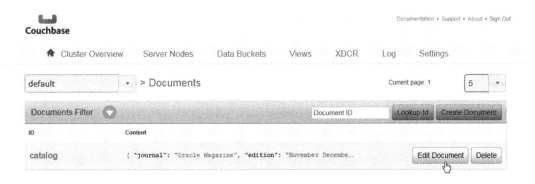

*Figure 9-15.* *Selecting Edit Document*

The document migrated from MongoDB is listed as a JSON document as shown in Figure 9-16.

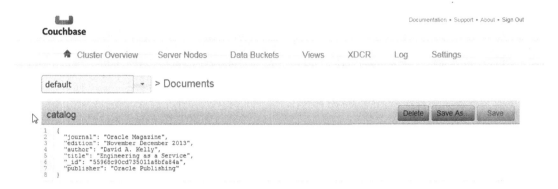

***Figure 9-16.*** *Couchbase Document migrated from MongoDB*

# Summary

MongoDB is an NoSQL database based on the JSON-like BSON format. In this chapter we discussed some of the advantages of Couchbase Server over MongoDB and subsequently migrated a document stored in MongoDB to Couchbase. First, we created a BSON document in MongoDB using the Mongo DB Client Java driver, and subsequently we used the Mongo DB Java driver to migrate the MongoDB document to Couchbase. In the next chapter we shall migrate Apache Cassandra, another NoSQL database, to Couchbase Server.

■ ■ ■

# CHAPTER 10

# Migrating Apache Cassandra

Apache Cassandra is a NoSQL, highly available, distributed database based on a row/column structure. The top-level namespace in Cassandra is *Keyspace*. A Keyspace is the equivalent of a database instance in an SQL relational database. An installation of Cassandra may have several Keyspaces. The top-level data structure for data storage is *Column Family* (also called a table), which is a set of *key-value pairs*. A Column Family definition consists of columns with one of the columns being the primary key column and the other columns being the data columns. A *Column* is the smallest unit of data stored in Cassandra and is associated with a name, a value, and a timestamp. One of the columns in a Column Family is the primary key (or row key). A primary key is identified with PRIMARY KEY in a column family definition.

Cassandra is not based on the JSON document object model, which has built-in support for hierarchical structures. In this chapter we shall migrate rows of data stored in Cassandra to Couchbase. Though any Cassandra client may be used to retrieve data from Cassandra and a Couchbase client in the same language and may be used to store the Cassandra data in Couchbase, we shall use a Java client for Cassandra and a Java client for Couchbase. The Java client for Cassandra that we shall use is the Datastax Java driver. The Java client for Couchbase that we shall use is the Couchbase Java Client. This chapter covers the following topics.

- Setting Up the Environment
- Creating a Maven Project in Eclipse
- Creating a Database in Cassandra
- Migrating the Cassandra Database to Couchbase

## Setting Up the Environment

Download the following software for Apache Cassandra and Couchbase Server.

- Couchbase Server Enterprise Edition 3.0.3 couchbase-server-enterprise_3.0.3-windows_amd64.exe file from http://www.couchbase.com/download. Double-click on the exe file to launch the installer and install Couchbase Server.

- Eclipse IDE for Java EE Developers from http://www.eclipse.org/downloads/.

- Apache Cassandra 2.1.7 apache-cassandra-2.1.7-bin.tar.gz from http://cassandra.apache.org/download/. Extract tar.gz file to a directory and add the bin directory, for example the C:\apache-cassandra-2.1.7\bin directory to the PATH variable.

Start Apache Cassandra server with the following command.

```
cassandra -f
```

The server gets started as shown in the server output as shown in Figure 10-1.

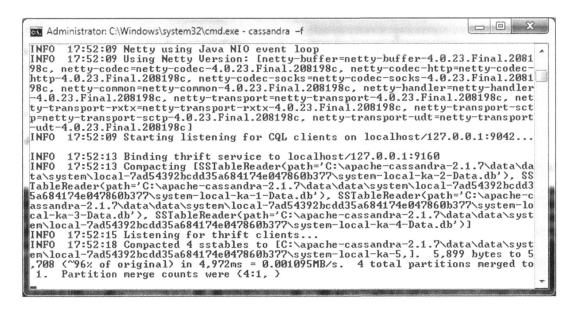

*Figure 10-1.* *Starting Apache Cassandra*

# Creating a Maven Project in Eclipse

Next, create a Java project in Eclipse IDE for migrating Cassandra database data to a Couchbase database.

1. Select File ➤ New ➤ Other.

2. In the New window, select Maven ➤ Maven Project and click on Next as shown in Figure 10-2.

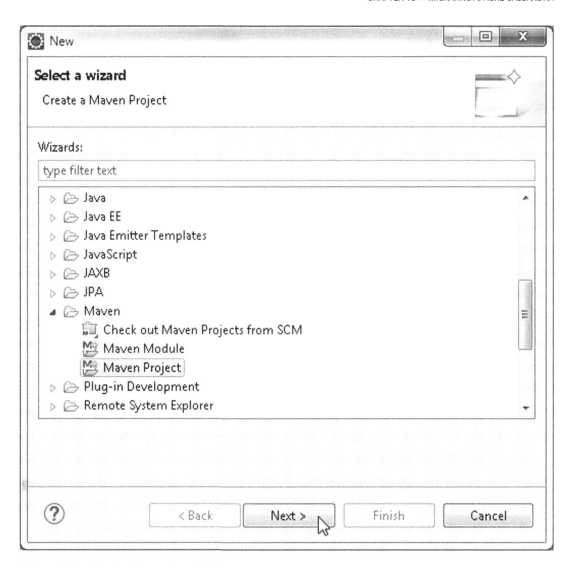

**Figure 10-2.** *Selecting Java ➤ Java Project*

3. The New Maven Project wizard gets started. Select the Create a simple project check box and the Use default Workspace location check box and click on Next as shown in Figure 10-3.

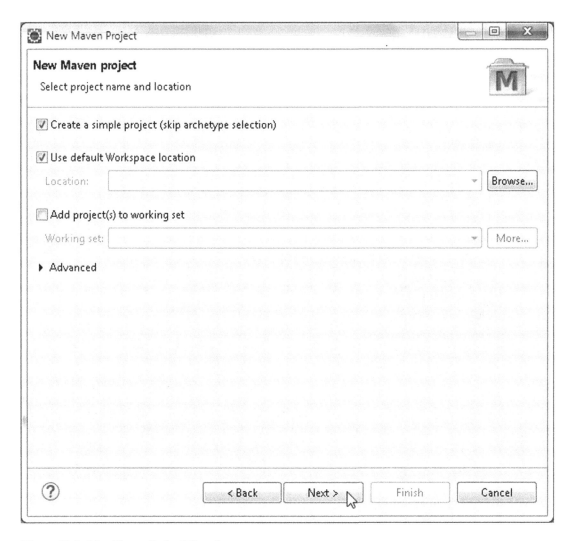

*Figure 10-3.* *New Maven Project Wizard*

4. In Configure project, specify the following and then click on Finish as shown in Figure 10-4.

- Group Id: `com.couchbase.migration`

- Artifact Id: `CassandraToCouchbase`

- Version: 1.0.0

- Packaging: jar

- Name: `CassandraToCouchbase`

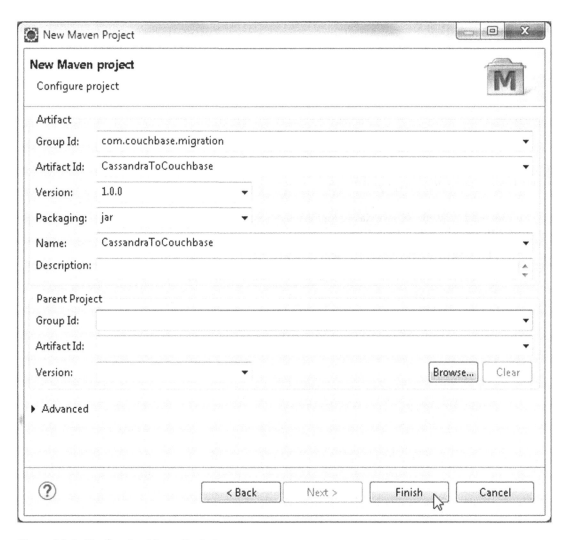

***Figure 10-4.*** *Configuring Maven Project*

A Maven Project gets created in Eclipse IDE as shown in Figure 10-5.

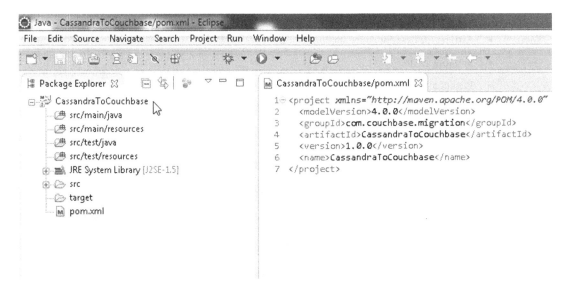

**Figure 10-5.** *Maven Project CassandraToCouchbase in Package Explorer*

Now we need to create two Java classes for the migration: one to create the initial data in Cassandra and the other to migrate the data to Couchbase.

1. To create a Java class click on File ➤ New ➤ Other.

2. In New select Java ➤ Class and click on Next as shown in Figure 10-6.

*Figure 10-6. Selecting Java ➤ Java Class*

3. In New Java Class wizard select the Source folder as src and specify Package as couchbase and class name as CreateCassandraDatabase. Click on Finish as shown in Figure 10-7.

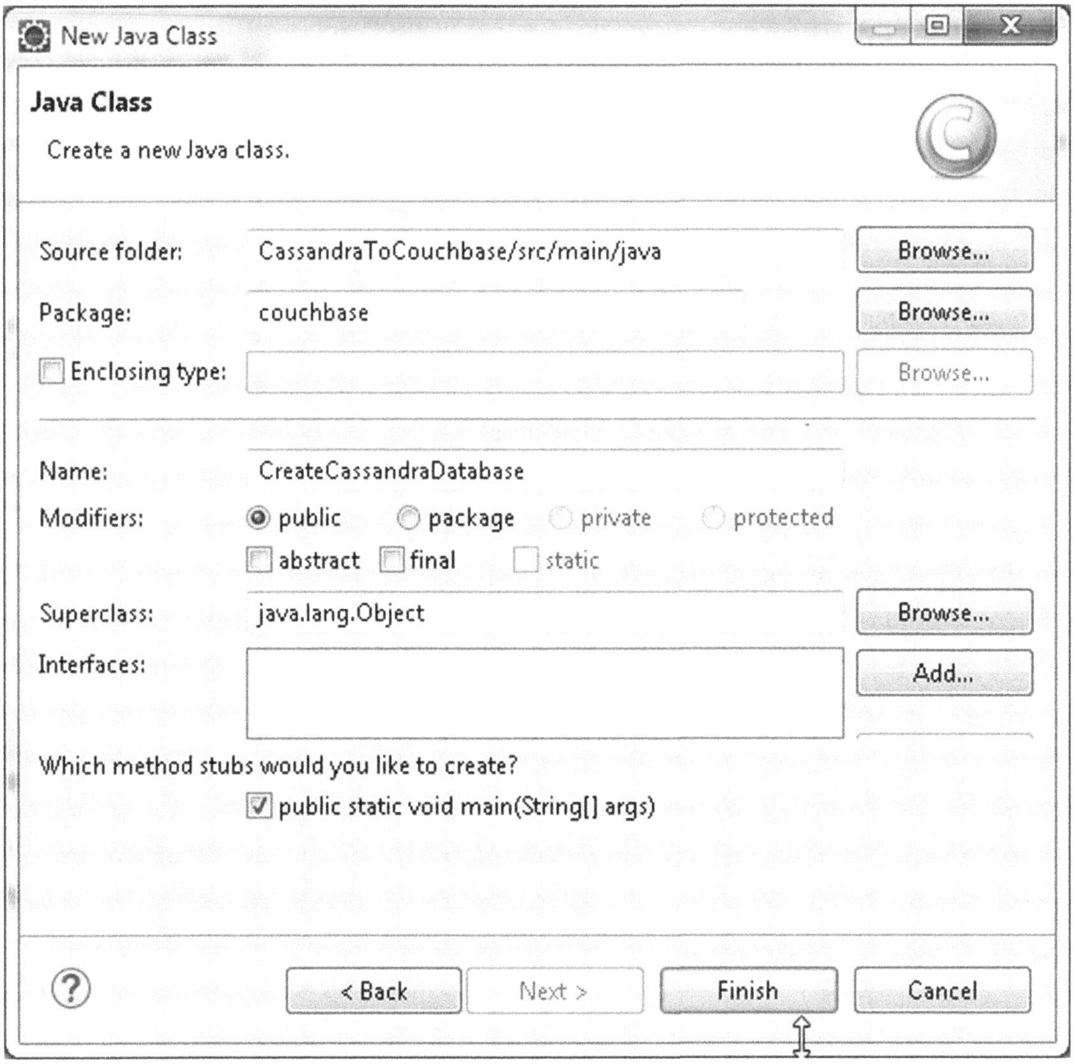

*Figure 10-7. Configuring Java Class CreateCassandraDatabase*

4. Similarly create a Java class `MigrateCassandraToCouchbase` as shown in Figure 10-8.

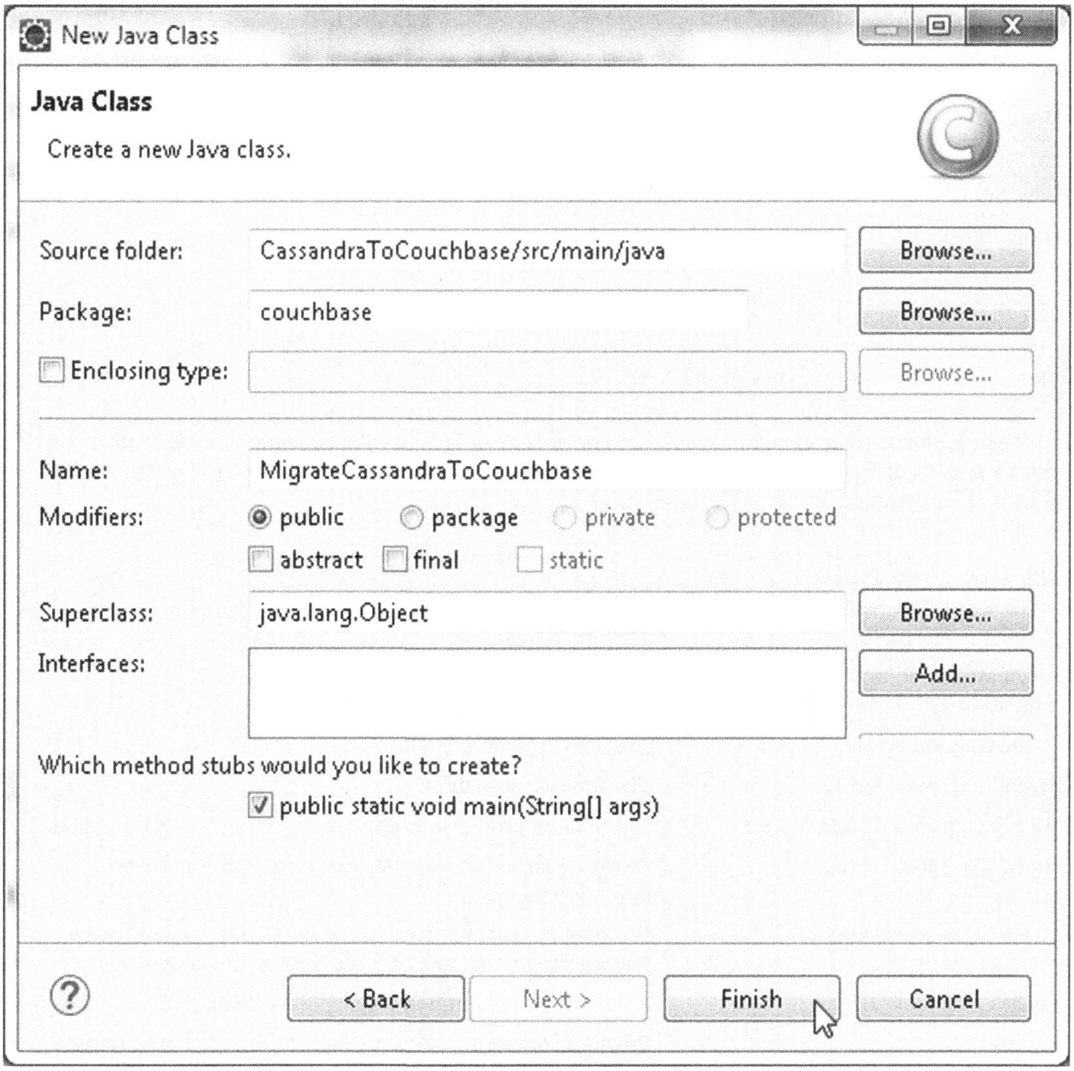

**Figure 10-8.** *Configuring Java Class MigrateCassandraToCouchbase*

The two Java classes are shown in the Package Explorer in Figure 10-9.

**Figure 10-9.** *Java Classes in Package Explorer*

We need to add some dependencies to the pom.xml. Add the following dependencies listed in Table 10-1; some of the dependencies are indicated as being included with the Apache Cassandra Project dependency and should not be added separately.

**Table 10-1.** *Dependencies*

| Jar | Description |
|-----|-------------|
| Couchbase server Java SDK Client library 2.1.3 | The Java Client to Couchbase Server. |
| Apache Cassandra | Apache Cassandra Project. |
| Cassandra Driver Core | The Datastax Java driver. |
| Apache Commons BeanUtils 1.9.2 | Utility Jar for Java classes developed with the JavaBeans pattern. |
| Apache Commons Collections | Provides Data Structures that accelerate Java application development. |
| Apache Commons Lang3 | Provides extra classes for manipulation of Java core classes. Included with Apache Cassandra Project dependency. |
| Apache Commons Logging | An interface for common logging implementations. |
| EZMorph | Provides conversion from one object to another and is used to convert between non-JSON objects and JSON objects. |
| Guava | Google's core libraries used in Java-based projects. Included with Apache Cassandra dependency. |
| Jackson Core Asl | High performance JSON processor. Included with Apache Cassandra Project dependency. |
| Jackson Mapper Asl | High performance data binding a package built on Jackson JSON processor. Included with Apache Cassandra Project dependency. |

(*continued*)

**Table 10-1.** (*continued*)

| Jar | Description |
| --- | --- |
| Metrics Core | The core library for Metrics. Included with Apache Cassandra Project dependency. |
| Netty | NIO client server framework to develop network applications such as protocol servers and clients. Included with Apache Cassandra Project dependency. |
| Slf4j Api | Simple Logging Façade for Java, which serves as an abstraction for various logging frameworks. Included with Apache Cassandra Project dependency. |

The pom.xml is listed below.

```xml
<project xmlns="http://maven.apache.org/POM/4.0.0" xmlns:xsi="http://www.w3.org/2001/
XMLSchema-instance"
    xsi:schemaLocation="http://maven.apache.org/POM/4.0.0 http://maven.apache.org/xsd/
    maven-4.0.0.xsd">
    <modelVersion>4.0.0</modelVersion>
    <groupId>com.couchbase.migration</groupId>
    <artifactId>CassandraToCouchbase</artifactId>
    <version>1.0.0</version>
    <name>CassandraToCouchbase</name>
    <dependencies>
        <dependency>
            <groupId>com.couchbase.client</groupId>
            <artifactId>java-client</artifactId>
            <version>2.1.4</version>
        </dependency>
        <dependency>
            <groupId>com.datastax.cassandra</groupId>
            <artifactId>cassandra-driver-core</artifactId>
            <version>2.1.6</version>
        </dependency>
        <dependency>
            <groupId>org.apache.cassandra</groupId>
            <artifactId>cassandra-all</artifactId>
            <version>2.1.7</version>
        </dependency>
        <dependency>
            <groupId>commons-beanutils</groupId>
            <artifactId>commons-beanutils</artifactId>
            <version>1.9.2</version>
        </dependency>
        <dependency>
            <groupId>commons-collections</groupId>
            <artifactId>commons-collections</artifactId>
            <version>3.2.1</version>
        </dependency>
```

```
        <dependency>
            <groupId>commons-logging</groupId>
            <artifactId>commons-logging</artifactId>
            <version>1.2</version>
        </dependency>
        <dependency>
            <groupId>net.sf.ezmorph</groupId>
            <artifactId>ezmorph</artifactId>
            <version>1.0.6</version>
        </dependency>
    </dependencies>
</project>
```

Some of these dependencies have further dependencies, which get added automatically and should not be added separately. To find the required Jars that get added from the dependencies, right-click on the project node in Package Explorer and select Properties. In Properties select Java Build Path. The Jars added to the migration project are shown in Figure 10-10.

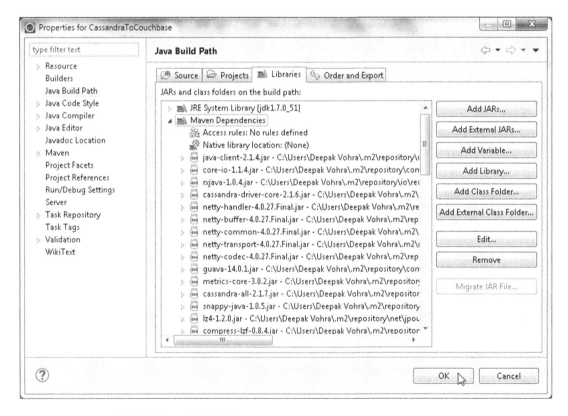

*Figure 10-10.* *Jar Files in the Java Build Path*

# Creating a Database in Cassandra

Before we are able to migrate Cassandra data to Couchbase we must create Cassandra data. Cassandra table may be created either using the Cassandra-Cli or using a Java application with Cassandra Java driver. We shall create Cassandra data in a Java application. We shall use the CreateCassandraDatabase class for creating a Cassandra database. First, we need to connect to Cassandra from the application. Create an instance of Cluster, which is the main entry point for the Datastax Java driver. The Cluster maintains a connection with one of the server nodes to keep information on the state and current topology of the cluster. The driver discovers all the nodes in the cluster using auto-discovery of nodes including new nodes that join later. Build a Cluster.Builder instance, which is a helper class to build Cluster instances, using static method builder().

We need to provide the connect address of at least one of the nodes in the Cassandra cluster for the Datastax driver to be able to connect with the cluster and discover other nodes in the cluster using auto-discovery. Using the addContactPoint(String) method of Cluster.Builder, add the address of the Cassandra server running on the localhost (127.0.0.1). Next, invoke the build() method to build the Cluster using the configured address/es. The methods may be invoked in sequence as we don't need the intermediary Cluster.Builder instance.

```
cluster = Cluster.builder().addContactPoint("127.0.0.1").build();
```

Next, invoke the connect() method to create a session on the cluster. A session is represented with the Session class, which holds multiple connections to the cluster. A Session instance is used to query the cluster. The Session instance provides policies on which node in the cluster to use for querying the cluster. The default policy is to use a round-robin on all the nodes in the cluster. Session is also used to handle retries of failed queries. Session instances are thread-safe and a single instance is sufficient for an application. But, a separate Session instance is required if connecting to multiple keyspaces as a single Session instance is specific to a particular keyspace only.

```
Session session = cluster.connect();
```

The Cassandra server must be running to be able to connect to the server when the application is run. If Cassandra server is not running the following exception is generated when a connection is tried.

```
com.datastax.driver.core.exceptions.NoHostAvailableException: All host(s) tried for query failed
(tried: /127.0.0.1 (com.datastax.driver.core.TransportException: [/127.0.0.1] Cannot connect))
    at com.datastax.driver.core.ControlConnection.reconnectInternal(ControlConnection.java:179)
    at com.datastax.driver.core.ControlConnection.connect(ControlConnection.java:77)
    at com.datastax.driver.core.Cluster$Manager.init(Cluster.java:890)
    at com.datastax.driver.core.Cluster$Manager.access$100(Cluster.java:806)
    at com.datastax.driver.core.Cluster.getMetadata(Cluster.java:217)
    at datastax.CQLClient.connection(CQLClient.java:43)
    at datastax.CQLClient.main(CQLClient.java:23)
```

The Session class provides several methods to prepare and run queries on the server, some of which are discussed in Table 10-2.

*Table 10-2.* *Session Class Methods to run Queries*

| Method | Description |
|--------|-------------|
| execute(Statement statement) | Executes the query provided as a Statement object to return a ResultSet. |
| execute(String query) | Executes the query provided as a String to return a ResultSet. |
| execute(String query, Object... values) | Executes the query provided as a String and using the specified values to return a ResultSet. |

We need to create a keyspace to store tables in. Add a method createKeyspace() to create a keyspace to the CreateCassandraDatabase application. CQL 3 (Cassandra Query Language 3) has added support to run CREATE statements conditionally, which is only if the object to be constructed does not already exist. The IF NOT EXISTS clause is used to create conditionally. Create a keyspace called datastax using replication with strategy class as SimpleStrategy and replication factor as 1.

```
session.execute("CREATE KEYSPACE IF NOT EXISTS datastax WITH replication "
+ "= {'class':'SimpleStrategy', 'replication_factor':1};");
```

Invoke the createKeyspace() method in the main method. When the application is run, a keyspace gets created. Cassandra supports the following strategy classes listed in Table 10-3 that refer to the replica placement strategy class.

*Table 10-3.* *Strategy Classes*

| Class | Description |
|-------|-------------|
| org.apache.cassandra. locator.SimpleStrategy | Used for a single data center only. The first replica is placed on a node as determined by the partitioner. Subsequent replicas are placed on the next node/s in a clockwise manner in the ring of nodes without consideration to topology. The replication factor is required only if SimpleStrategy class is used. |
| org.apache.cassandra.locator. NetworkTopologyStrategy | Used with multiple data centers. Specifies how many replicas to store in each data center. Attempts to store replicas on different racks within the same data center because nodes in the same rack are more likely to fail together. |

Next, we shall create a column family, which is also called a table in CQL 3. Add a method createTable() to CreateCassandraDatabase. CREATE TABLE command also supports IF NOT EXISTS to create a table conditionally. CQL 3 has added the provision to create a compound primary key, a primary key created from multiple component primary key columns. In a compound primary key the first column is called the partition key. Create a table called catalog, which has columns catalog_id, journal, publisher, edition, title and author. In catalog table the compound primary key is made from catalog_id, journal columns with catalog_id being the partition key. Invoke the execute(String) method to create table catalog as follows.

```
session.execute("CREATE TABLE IF NOT EXISTS datastax.catalog (catalog_id text,journal text,
publisher text, edition text,title text,author text,PRIMARY KEY (catalog_id, journal))");
```

Prefix the table name with the keyspace name. Invoke the createTable method in main method. When the CreateCassandraDatabase application is run the catalog table gets created. Next, we shall add data to the table catalog using the INSERT statement. Use the IF NOT EXISTS keyword to add rows conditionally. When a compound primary key is used all the component primary key columns must be specified, including the values for the compound key columns.

Add a method insert() to the CreateCassandraDatabase class and invoke the method in the main method. Add two rows identified by row ids catalog1, catalog2 to the table catalog. For example, the two rows are added to the catalog table as follows.

```
session.execute("INSERT INTO datastax.catalog (catalog_id, journal, publisher,
edition,title,author) VALUES ('catalog1','Oracle Magazine', 'Oracle Publishing',
'November-December 2013', 'Engineering as a Service','David A.  Kelly') IF NOT EXISTS");
session.execute("INSERT INTO datastax.catalog (catalog_id, journal, publisher,
edition,title,author) VALUES ('catalog2','Oracle Magazine', 'Oracle Publishing',
'November-December 2013', 'Quintessential and Collaborative','Tom Haunert') IF NOT EXISTS");
```

Next, we shall run a SELECT statement to select columns from the catalog table. Add a method select() to run SELECT statement/s. Select all the columns from the catalog table using the * for column selection. The SELECT statement is run as a test to find that the data we added got added.

```
ResultSet results = session.execute("select * from datastax.catalog");
```

A row in the ResultSet is represented with the Row class. Iterate over the ResultSet to output the column value or each of the columns.

```
for (Row row : results) {

        System.out.println("Journal: " + row.getString("journal"));
        System.out.println("Publisher: " + row.getString("publisher"));
        System.out.println("Edition: " + row.getString("edition"));
        System.out.println("Title: " + row.getString("title"));
        System.out.println("Author: " + row.getString("author"));
        System.out.println("\n");
        System.out.println("\n");
    }
```

The CreateCassandraDatabase class is listed below.

```
package couchbase;

import com.datastax.driver.core.Cluster;
import com.datastax.driver.core.ResultSet;
import com.datastax.driver.core.Row;
import com.datastax.driver.core.Session;

public class CreateCassandraDatabase {
    private static Cluster cluster;
    private static Session session;
    public static void main(String[] argv) {
        cluster = Cluster.builder().addContactPoint("127.0.0.1").build();
        session = cluster.connect();
        createKeyspace();
```

```java
        createTable();
        insert();
        select();
    }
    private static void createKeyspace() {
        session.execute("CREATE KEYSPACE IF NOT EXISTS datastax WITH replication "
                + "= {'class':'SimpleStrategy', 'replication_factor':1};");
    }
    private static void createTable() {
        session.execute("CREATE TABLE IF NOT EXISTS datastax.catalog (catalog_id text,journal
text,publisher text, edition text,title text,author text,PRIMARY KEY (catalog_id, journal))");
    }
    private static void insert() {
        session.execute("INSERT INTO datastax.catalog (catalog_id, journal, publisher,
edition,title,author) VALUES ('catalog1','Oracle Magazine', 'Oracle Publishing', 'November-
December 2013', 'Engineering as a Service','David A.  Kelly') IF NOT EXISTS");
        session.execute("INSERT INTO datastax.catalog (catalog_id, journal, publisher,
edition,title,author) VALUES ('catalog2','Oracle Magazine', 'Oracle Publishing', 'November-
December 2013', 'Quintessential and Collaborative','Tom Haunert') IF NOT EXISTS");
    }
    private static void select() {
        ResultSet results = session.execute("select * from datastax.catalog");
        for (Row row : results) {
            System.out.println("Catalog Id: " + row.getString("catalog_id"));
            System.out.println("\n");
            System.out.println("Journal: " + row.getString("journal"));
            System.out.println("\n");
            System.out.println("Publisher: " + row.getString("publisher"));
            System.out.println("\n");
            System.out.println("Edition: " + row.getString("edition"));
            System.out.println("\n");
            System.out.println("Title: " + row.getString("title"));
            System.out.println("\n");
            System.out.println("Author: " + row.getString("author"));
            System.out.println("\n");
        }
    }
}
```

Run the CreateCassandraDatabase application to add two rows of data to the catalog table. Right-click on CreateCassandraDatabase.java in Package Explorer and select Run As ➤ Java Application as shown in Figure 10-11.

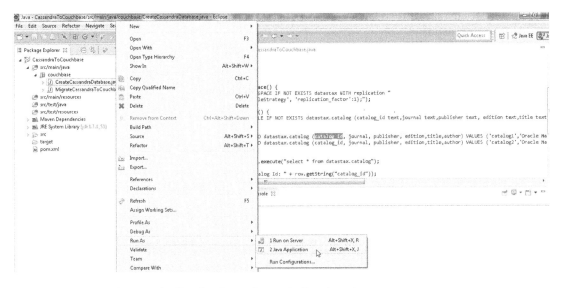

***Figure 10-11.*** *Running Java Application CreateCassandraDatabase.java*

The Cassandra keyspace datastax gets created, the catalog table gets created, and data gets added to the table. The SELECT statement, which is run as a test, outputs the two rows added to Cassandra as shown in Figure 10-12.

***Figure 10-12.*** *The two rows of data added to Cassandra Database*

To verify that the datastax keyspace got created in Cassandra, log in to the Cassandra Client interface with the following command.

```
cassandra-cli
```

Run the following command to authenticate the datastax keyspace.

```
use datastax;
```

The datastax keyspace gets authenticated as shown in Figure 10-13.

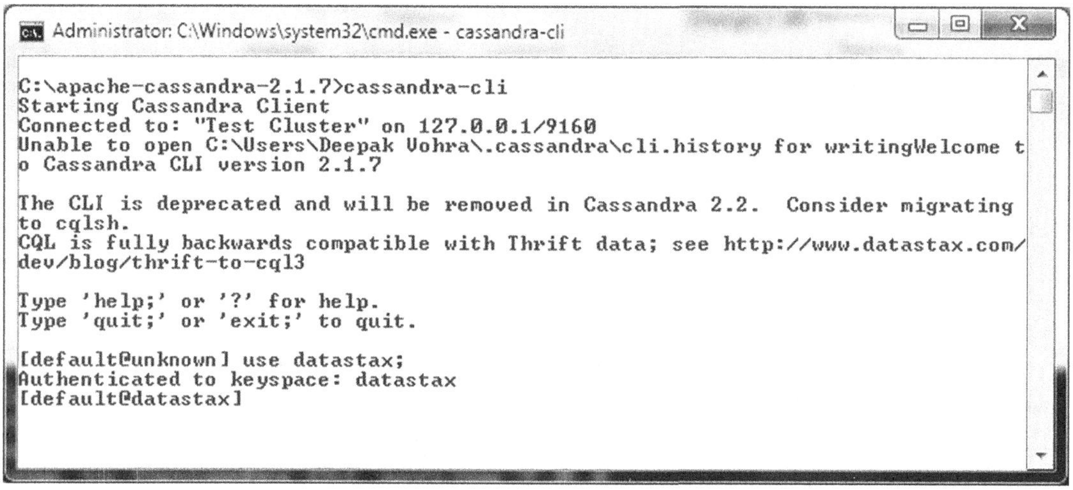

**Figure 10-13.** *Selecting the datastax keyspace*

To output the table stored in Cassandra run the following commands in Cassandra-Cli.

```
assume catalog keys as utf8;
assume catalog validator as utf8;
assume catalog comparator as utf8;
GET catalog[utf8('catalog1')];
GET catalog[utf8('catalog2')];
```

The two rows stored in the catalog table get listed as shown in Figure 10-14.

```
Administrator: C:\Windows\system32\cmd.exe - cassandra-cli
[default@datastax]
[default@datastax] assume catalog keys as utf8;
Assumption for column family 'catalog' added successfully.
[default@datastax] assume catalog validator as utf8;
Assumption for column family 'catalog' added successfully.
[default@datastax] assume catalog comparator as utf8;
Assumption for column family 'catalog' added successfully.
[default@datastax] GET catalog[utf8('catalog1')];
=> (name= *Oracle Magazine    , value=, timestamp=1436468582885000)
=> (name= *Oracle Magazine  +author , value=David A.  Kelly, timestamp=143646858
2885000)
=> (name= *Oracle Magazine  edition , value=November-December 2013, timestamp=14
36468582885000)
=> (name= *Oracle Magazine       publisher , value=Oracle Publishing, timestamp=1
436468582885000)
=> (name= *Oracle Magazine  +title , value=Engineering as a Service, timestamp=1
436468582885000)
Returned 5 results.
Elapsed time: 36 msec(s).
[default@datastax] GET catalog[utf8('catalog2')];
=> (name= *Oracle Magazine    , value=, timestamp=1436468583077000)
=> (name= *Oracle Magazine  +author , value=Tom Haunert, timestamp=1436468583077
000)
=> (name= *Oracle Magazine  edition , value=November-December 2013, timestamp=14
36468583077000)
=> (name= *Oracle Magazine       publisher , value=Oracle Publishing, timestamp=1
436468583077000)
=> (name= *Oracle Magazine  +title , value=Quintessential and Collaborative, tim
estamp=1436468583077000)
Returned 5 results.
Elapsed time: 65 msec(s).
[default@datastax] _
```

***Figure 10-14.*** *Listing the catalog table Rows*

Next, we shall migrate the Cassandra data to Couchbase Server.

# Migrating the Cassandra Database to Couchbase

In this section we shall get the data stored earlier in Cassandra NoSQL database and migrate the data to Couchbase Server. We shall use the MigrateCassandraToCouchbase class to migrate the data from Cassandra database to Couchbase Server. Add a method called migrate() to the MigrateCassandraToCouchbase class and invoke the method from the main method. From the MigrateCassandraToCouchbase class, connect to the Cassandra server as explained in the previous section in the main method.

```
cluster = Cluster.builder().addContactPoint("127.0.0.1").build();
session = cluster.connect();
```

A Session object is created to represent a connection with Cassandra server. We shall use the Session object to run a SELECT statement on Cassandra to select the data to be migrated. Run a SELECT statement as follows to select all rows from the catalog table in the datastax keyspace in the migrate() method.

```
ResultSet results = session.execute("select * from datastax.catalog");
```

The result set of the query is represented with the ResultSet class. A row in the ResultSet is represented with the Row class. Iterate over the ResultSet to fetch each row as a Row object.

```
for (Row row : results) {
}
```

Before we migrate the rows of data fetched from Cassandra create a Java client for Couchbase because we would need to add the fetched data to Couchbase. As discussed in Chapter 2, the CouchbaseCluster class is the main entry point for connecting to the Couchbase Server. In the migrate() method create a CouchbaseCluster instance using the static method create().

```
Cluster cluster = CouchbaseCluster.create();
```

Couchbase Server stores documents in Data Buckets. The default data bucket is called "default." As discussed in Chapter 2, the CouchbaseCluster provides the overloaded openBucket() method to connect to a Couchbase bucket. Create a Bucket instance for the default bucket.

```
Bucket defaultBucket = cluster.openBucket("default");
```

Couchbase server stores documents as JSON. Next, we shall construct JSON strings from data retrieved for Cassandra and store the JSON string in Couchbase. In the migrate() method add a counter variable for the rows fetched from Cassandra. Next, we shall migrate the data using a for loop row by row. Add an int counter i for the rows of data and increment the counter in the for loop with each row iterated.

```
int i = 0;
    for (Row row : results) {
    i = i + 1;
}
```

An unordered collection of name-value pairs that constitute a JSON document is represented by the com.couchbase.client.java.document.json.JsonObject class. Within the for loop create an empty JsonObject instance, which represents a JSON object that may be stored in Couchbase Server, using the static method empty().

```
JsonObject catalogObj = JsonObject.empty();
```

Obtain the column definitions as represented by a ColumnDefinitions object using the getColumnDefinitions() method of Row. Create an Iterator over the ColumnDefinitions object using the iterator() method with each column definition being represented with ColumnDefinitions.Definition. Using the Iterator in conjunction with its hasNext method iterate over the columns and obtain the column names.

```
while (iter.hasNext()) {
    ColumnDefinitions.Definition column = iter.next();
    String columnName = column.getName();
}
```

Using the getString(String columnName) method in Row obtain the value corresponding to each column.

```
String value = row.getString(columnName);
```

Using the put(String key, String value) method in JsonObject put the column name and column value in the JsonObject object.

```
catalogObj.put(columnName, value);
```

The Bucket class provides overloaded insert and upsert methods to store/add documents to Couchbase Server. We shall use the insert(D document) method. Add the JsonObject instance created earlier to the default bucket using the Bucket instance and the insert(D document) method. Create an instance of JsonDocument to add using the insert(D document) method with the JsonDocument. create(java.lang. String id, JsonObject content) method. Specify document id as "catalog" suffixed with the row counter i.

```
JsonDocument document = defaultBucket.insert(JsonDocument.create("catalog" + i, catalogObj));
```

The MigrateCassandraToCouchbase class is listed below.

```
package couchbase;

import java.net.URI;
import java.util.Iterator;
import java.util.LinkedList;
import java.util.List;

import com.couchbase.client.java.Bucket;
import com.couchbase.client.java.CouchbaseCluster;
import com.couchbase.client.java.Cluster;
import com.couchbase.client.java.document.JsonDocument;
import com.couchbase.client.java.document.json.JsonObject;
import com.datastax.driver.core.ColumnDefinitions;
import com.datastax.driver.core.ResultSet;
import com.datastax.driver.core.Row;
import com.datastax.driver.core.Session;

public class MigrateCassandraToCouchbase {
    private static com.datastax.driver.core.Cluster cluster;
    private static Session session;

    public static void main(String[] argv) {
        cluster = com.datastax.driver.core.Cluster.builder()
                .addContactPoint("127.0.0.1").build();
        session = cluster.connect();
        migrate();
    }

    private static void migrate() {
        Cluster cluster = CouchbaseCluster.create();
        Bucket defaultBucket = cluster.openBucket("default");
        ResultSet results = session.execute("select * from datastax.catalog");
        int i = 0;
        for (Row row : results) {
            i = i + 1;
            JsonObject catalogObj = JsonObject.empty();
```

```
        ColumnDefinitions columnDefinitions = row.getColumnDefinitions();
        Iterator<ColumnDefinitions.Definition> iter = columnDefinitions
                .iterator();
        while (iter.hasNext()) {
            ColumnDefinitions.Definition column = iter.next();
            String columnName = column.getName();
            String value = row.getString(columnName);
            catalogObj.put(columnName, value);

        }
        JsonDocument document = defaultBucket.insert(JsonDocument
                .create("catalog" + i, catalogObj));
        System.out.println("Set Succeeded");
    }
  }

}
```

Next, we shall run the MigrateCassandraToCouchbase application in Eclipse IDE. Right-click on MigrateCassandraToCouchbase.java and select Run As ➤ Java Application as shown in Figure 10-15.

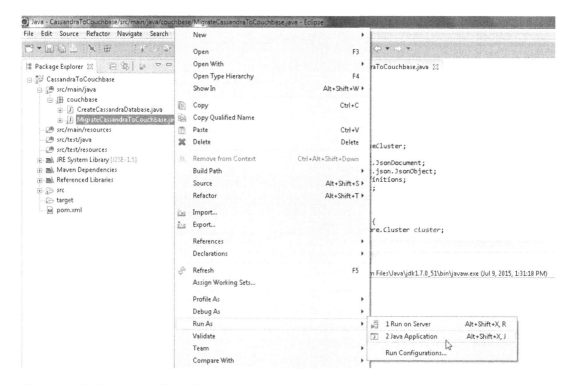

***Figure 10-15.*** *Running the MigrateCassandraToCouchbase.java Application*

Two rows of data get fetched from Cassandra and get stored in Couchbase Server. Now, complete the following steps.

1. Log in to the Couchbase Administration Console with URL
   `http://localhost:8091/index.html`.

2. Specify the Username and Password to log in to the Console and click on Sign In.

3. In the Console, click on Data Buckets.

4. The Couchbase Buckets gets listed, the "default" bucket being one of them.

5. The Item Count for the default bucket should be listed as 2 for the two documents migrated from Cassandra as shown in Figure 10-16.

***Figure 10-16.*** *Selecting the Documents button*

6. Click on the Documents button for the default bucket.

7. The two documents `catalog1` and `catalog2` are listed as added to the `default` bucket as shown in Figure 10-17. Click on the Edit Document button for a document to list the JSON document. For example, click on the Edit Document button for the `catalog1` document as shown in Figure 10-17.

***Figure 10-17.*** *The two documents migrated from Cassandra to Couchbase*

The catalog1 document JSON gets listed as shown in Figure 10-18.

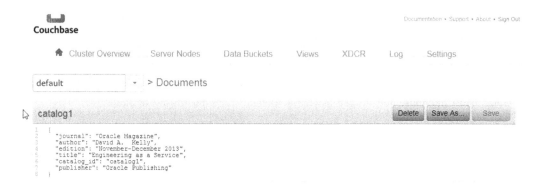

***Figure 10-18.*** *The catalog1 JSON Document*

8. Similarly, click on the Edit Document button for the catalog2 document as shown in Figure 10-19.

***Figure 10-19.*** *Selecting the catalog2 JSON document*

The `catalog2` id JSON document gets displayed as shown in Figure 10-20.

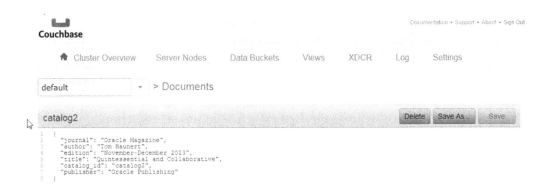

*Figure 10-20.* *catalog2 JSON*

We migrated two rows listed in Table 10-4 from Cassandra server to two JSON documents in Couchbase Server.

*Table 10-4.* *The two rows migrated from Cassandra to Couchbase*

| Row | catalog_id | journal | publisher | edition | title | author | JSON Document |
|---|---|---|---|---|---|---|---|
| Row 1 | catalog1 | Oracle Magazine | Oracle Publishing | November-December 2013 | Engineering as a Service | David A. Kelly | {<br>"catalog_id": "catalog1",<br>"journal": "Oracle Magazine",<br>"author": "David A. Kelly",<br>"edition": "November-December 2013",<br>"publisher": "Oracle Publishing",<br>"title": "Engineering as a Service"<br>} |

(*continued*)

***Table 10-4.*** (*continued*)

| Row | catalog_id | journal | publisher | edition | title | author | JSON Document |
|-----|-----------|---------|-----------|---------|-------|--------|---------------|
| Row 2 | catalog2 | Oracle Magazine | Oracle Publishing | November-December 2013 | Quintessential and Collaborative | Tom Haunert | {<br><br>"catalog_id": "catalog2",<br>"journal": "Oracle Magazine",<br>"author": "Tom Haunert",<br>"edition": "November-December 2013",<br>"publisher": "Oracle Publishing",<br>"title": "Quintessential and Collaborative"<br>} |

We did not hard-code any column names or column values. Any number of rows with each row having any columns and any number of columns may be migrated using the `MigrateCassandraToCouchbase` application. The keyspace and table name in the `SELECT` statement to fetch data from Cassandra are the only two values hard-coded and may be modified as required.

# Summary

In this chapter we migrated data from Cassandra database to Couchbase database using a Java application. First, we created sample data in Cassandra. Subsequently we fetched the sample data in a Java application and migrated the data as JSON documents to Couchbase Server. In the next chapter we shall migrate Oracle Database to Couchbase Server.

# CHAPTER 11

■ ■ ■

# Migrating Oracle Database

Couchbase Server is one of the leading NoSQL databases in the document store type databases category. Couchbase stores documents using the JSON data format, which is the most flexible of data models with provision to create hierarchies of data structures. In contrast, Oracle database stores data in a fixed schema table format. The database model of Couchbase is based on Document store while in Oracle Database data is organized as tables with two-dimensional matrices made of columns and rows. In this chapter we shall migrate an Oracle Database table to Couchbase Server. A direct migration tool is not available for migrating from Oracle to Couchbase. We shall first export Oracle Database table to a CSV file. Subsequently we shall import the CSV file data into Couchbase Server using the *cbtransfer* tool. This chapter covers the following topics.

- Overview of the cbtransfer Tool
- Setting the Environment
- Creating an Oracle Database Table
- Exporting Oracle Database Table to CSV File
- Transferring Data from CSV File to Couchbase

## Overview of the cbtransfer Tool

The cbtransfer tool is used to transfer data between two Couchbase clusters or between a file and a Couchbase cluster. The syntax for using the cbtransfer tool is as follows.

```
cbtransfer [options] source destination
```

For example to transfer between a source cluster and a destination cluster run the following command.

```
cbtransfer http://SOURCE:8091 http://DEST:8091
```

To transfer from a backup to a cluster run the following command.

```
cbtransfer /backups/backup-42 http://DEST:8091
```

To transfer from a cluster to a backup run the following command.

```
cbtransfer http://SOURCE:8091 /backups/backup-42
```

We need to transfer from a CSV file to a cluster for which the command has the following format.

```
cbtransfer c:/wlslog.csv  http://DEST:8091
```

Some of the options supported by cbtransfer are discussed in Table 11-1.

*Table 11-1.*  *Options supported by cbtransfer Tool*

| Option | Description |
| --- | --- |
| -b BUCKET_SOURCE, --bucket-source=BUCKET_SOURCE | Specifies the source bucket. |
| -B BUCKET_DESTINATION, --bucket-destination=BUCKET_DESTINATION | Specifies the destination bucket. |
| -u USERNAME, --username=USERNAME | Username |
| -p PASSWORD, --password=PASSWORD | Password |
| -t THREADS, --threads=THREADS | Number of concurrent threads performing the transfer. |
| -i ID, --id=ID | Specifies the vbucketID of the items to transfer. |
| -k KEY, --key=KEY | Specifies the regexp for the item keys for the items to transfer. |
| -n, --dry-run | A dry run does not actually transfer data but just performs a validation of the files, connectivity, and configuration. |

Extra configuration parameters may be set using -x. Some of the extra configuration parameters are discussed in Table 11-2.

*Table 11-2.*  *Extra Configuration Parameters supported by cbtransfer Tool*

| Configuration Parameters | Description |
| --- | --- |
| batch_max_bytes | Transfer this # of bytes per batch |
| batch_max_size | Transfer this # of documents per batch |
| max_retry | Max number of sequential retries if transfer fails |
| recv_min_bytes | Amount of bytes for every TCP/IP call transferred |
| report | Number of batches transferred before updating progress bar in console |
| report_full | Number of batches transferred before emitting progress information in console |
| uncompress | For value 1, restore data in uncompressed mode |

# Setting the Environment

We need to download the following software for this chapter.

- Oracle Database 12c. Download from http://www.oracle.com/technetwork/database/enterprise-edition/downloads/index-092322.html.

- Couchbase Server (Version: 3.0.x Enterprise Edition).

- The cbtransfer tool is installed with Couchbase Server and for Windows located in C:\Program Files\Couchbase\Server\bin\ directory.

As a reminder, when installing and configuring Couchbase Server in the CREATE DEFAULT BUCKET for the default bucket, select Bucket Type as Couchbase as shown in Figure 11-1.

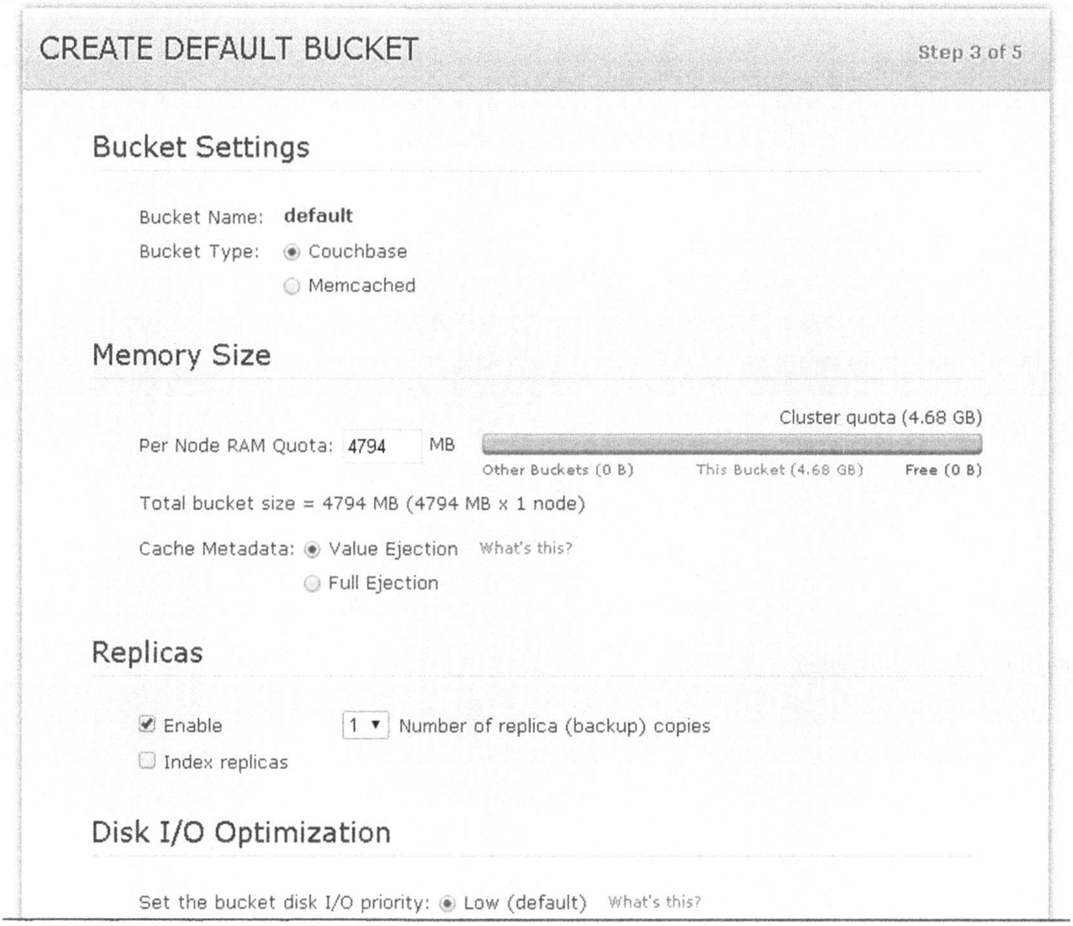

*Figure 11-1.* *Selecting Data Bucket Type as Couchbase*

Select the Enable checkbox for Flush as shown in Figure 11-2.

***Figure 11-2.*** *Enabling Flush*

# Creating an Oracle Database Table

First, create an Oracle Database table WLSLOG using the following SQL script.

```
CREATE TABLE OE.WLSLOG (ID VARCHAR2(255) PRIMARY KEY, TIME_STAMP VARCHAR2(255),
CATEGORY VARCHAR2(255), TYPE VARCHAR2(255), SERVERNAME VARCHAR2(255), CODE VARCHAR2(255),
MSG VARCHAR2(255));
INSERT INTO OE.WLSLOG (ID, TIME_STAMP, CATEGORY, TYPE, SERVERNAME, CODE, MSG) values
('catalog1','Apr-8-2014-7:06:16-PM-PDT','Notice','WebLogicServer','AdminServer',
'BEA-000365','Server state changed to STANDBY');
INSERT INTO OE.WLSLOG (ID, TIME_STAMP, CATEGORY, TYPE, SERVERNAME, CODE, MSG) values
('catalog2','Apr-8-2014-7:06:17-PM-PDT','Notice','WebLogicServer','AdminServer',
'BEA-000365','Server state changed to STARTING');
INSERT INTO OE.WLSLOG (ID,TIME_STAMP, CATEGORY, TYPE, SERVERNAME, CODE, MSG) values
('catalog3','Apr-8-2014-7:06:18-PM-PDT', 'Notice', 'WebLogicServer', 'AdminServer',
'BEA-000365', 'Server state changed to ADMIN');
INSERT INTO OE.WLSLOG (ID,TIME_STAMP, CATEGORY, TYPE, SERVERNAME, CODE, MSG) values
('catalog4','Apr-8-2014-7:06:19-PM-PDT', 'Notice', 'WebLogicServer', 'AdminServer',
'BEA-000365', 'Server state changed to RESUMING');
INSERT INTO OE.WLSLOG (ID,TIME_STAMP, CATEGORY, TYPE, SERVERNAME, CODE, MSG) values
('catalog5','Apr-8-2014-7:06:20-PM-PDT', 'Notice', 'WebLogicServer', 'AdminServer',
'BEA-000361', 'Started WebLogic AdminServer');
INSERT INTO OE.WLSLOG (ID,TIME_STAMP, CATEGORY, TYPE, SERVERNAME, CODE, MSG) values
('catalog6','Apr-8-2014-7:06:21-PM-PDT', 'Notice', 'WebLogicServer', 'AdminServer',
'BEA-000365', 'Server state changed to RUNNING');
INSERT INTO OE.WLSLOG (ID,TIME_STAMP, CATEGORY, TYPE, SERVERNAME, CODE, MSG) values
('catalog7','Apr-8-2014-7:06:22-PM-PDT', 'Notice', 'WebLogicServer', 'AdminServer',
'BEA-000360', 'Server started in RUNNING mode');
```

Run the SQL script in SQL*Plus to create the OE.WLSLOG table and add data to the table as shown in Figure 11-3.

*Figure 11-3.* *Creating Oracle Database Table OE.WLSLOG*

# Exporting Oracle Database Table to CSV File

Next, run the following SQL script in SQL*Plus to select data from the OE.WLSLOG table and export to a wlslog.csv file.

```
set pagesize 0 linesize 500 trimspool on feedback off echo off
select ID || ',' || TIME_STAMP || ',' || CATEGORY || ',' || TYPE || ',' || SERVERNAME || ','
|| CODE || ',' || MSG from OE.WLSLOG;
spool wlslog.csv
/
spool off
```

When the SQL script is run as shown in Figure 11-4, data is exported to the wlslog.csv file.

```
SQL> set pagesize 0 linesize 500 trimspool on feedback off echo off
SQL> select ID || ',' || TIME_STAMP || ',' || CATEGORY || ',' || TYPE || ',' ||
SERVERNAME || ',' || CODE || ',' || MSG from OE.WLSLOG;
catalog1,Apr-8-2014-7:06:16-PM-PDT,Notice,WebLogicServer,AdminServer,BEA-000365,
Server state changed to STANDBY
catalog2,Apr-8-2014-7:06:17-PM-PDT,Notice,WebLogicServer,AdminServer,BEA-000365,
Server state changed to STARTING
catalog3,Apr-8-2014-7:06:18-PM-PDT,Notice,WebLogicServer,AdminServer,BEA-000365,
Server state changed to ADMIN
catalog4,Apr-8-2014-7:06:19-PM-PDT,Notice,WebLogicServer,AdminServer,BEA-000365,
Server state changed to RESUMING
catalog5,Apr-8-2014-7:06:20-PM-PDT,Notice,WebLogicServer,AdminServer,BEA-000361,
Started WebLogic AdminServer
catalog6,Apr-8-2014-7:06:21-PM-PDT,Notice,WebLogicServer,AdminServer,BEA-000365,
Server state changed to RUNNING
catalog7,Apr-8-2014-7:06:22-PM-PDT,Notice,WebLogicServer,AdminServer,BEA-000360,
Server started in RUNNING mode
SQL> spool wlslog.csv
SQL> /
catalog1,Apr-8-2014-7:06:16-PM-PDT,Notice,WebLogicServer,AdminServer,BEA-000365,
Server state changed to STANDBY
catalog2,Apr-8-2014-7:06:17-PM-PDT,Notice,WebLogicServer,AdminServer,BEA-000365,
Server state changed to STARTING
catalog3,Apr-8-2014-7:06:18-PM-PDT,Notice,WebLogicServer,AdminServer,BEA-000365,
Server state changed to ADMIN
catalog4,Apr-8-2014-7:06:19-PM-PDT,Notice,WebLogicServer,AdminServer,BEA-000365,
Server state changed to RESUMING
catalog5,Apr-8-2014-7:06:20-PM-PDT,Notice,WebLogicServer,AdminServer,BEA-000361,
Started WebLogic AdminServer
catalog6,Apr-8-2014-7:06:21-PM-PDT,Notice,WebLogicServer,AdminServer,BEA-000365,
Server state changed to RUNNING
catalog7,Apr-8-2014-7:06:22-PM-PDT,Notice,WebLogicServer,AdminServer,BEA-000360,
Server started in RUNNING mode
SQL> spool off
SQL>
```

***Figure 11-4.*** *Exporting Oracle Database Table to CSV File*

Remove the leading and trailing output that is not the data exported to save the following as the wlslog.csv file.

```
catalog1,Apr-8-2014-7:06:16-PM-PDT,Notice,WebLogicServer,AdminServer,BEA-000365,Server state
changed to STANDBY
catalog2,Apr-8-2014-7:06:17-PM-PDT,Notice,WebLogicServer,AdminServer,BEA-000365,Server state
changed to STARTING
catalog3,Apr-8-2014-7:06:18-PM-PDT,Notice,WebLogicServer,AdminServer,BEA-000365,Server state
changed to ADMIN
```

```
catalog4,Apr-8-2014-7:06:19-PM-PDT,Notice,WebLogicServer,AdminServer,BEA-000365,Server state
changed to RESUMING
catalog5,Apr-8-2014-7:06:20-PM-PDT,Notice,WebLogicServer,AdminServer,BEA-000361,Started
WebLogic AdminServer
catalog6,Apr-8-2014-7:06:21-PM-PDT,Notice,WebLogicServer,AdminServer,BEA-000365,Server state
changed to RUNNING
catalog7,Apr-8-2014-7:06:22-PM-PDT,Notice,WebLogicServer,AdminServer,BEA-000360,Server
started in RUNNING mode
```

# Transferring Data from CSV File to Couchbase

In this section we shall transfer data from the wlslog.csv file to Couchbase Server using the cbtransfer tool. Log in to the Couchbase Admin Console with the URL localhost:8091. Click on Data Buckets. The default bucket should get listed with Item Count as 0 as shown in Figure 11-5.

***Figure 11-5.*** *The default bucket with Item Count as 0*

Run the cbtransfer tool to transfer data from the wlslog.csv file to the Couchbase Server default bucket. The command parameters in order are listed in Table 11-3.

***Table 11-3.*** *Command Parameters used for the cbtransfer Tool command*

| Parameter | Description |
|---|---|
| C:\Couchbase\wlslog.csv | The source CSV file to transfer data from |
| http://127.0.0.1:8091 | The destination Couchbase Server URL |
| -B default | The bucket to transfer data to |
| -u Administrator | The username |
| -p couchbase | The password |
| -x  batch_max_size | The # of documents to transfer per batch |
| -x max_retry=10 | The max number of sequential retries if transfer fails |

Run the following cbtransfer command.

```
cbtransfer C:\Couchbase\wlslog.csv http://127.0.0.1:8091 -B default -u Administrator -p
couchbase  -x  batch_max_size=2   -x max_retry=10
```

As the output indicates, data gets transferred to Couchbase Server from the wlslog.csv file as shown in Figure 11-6.

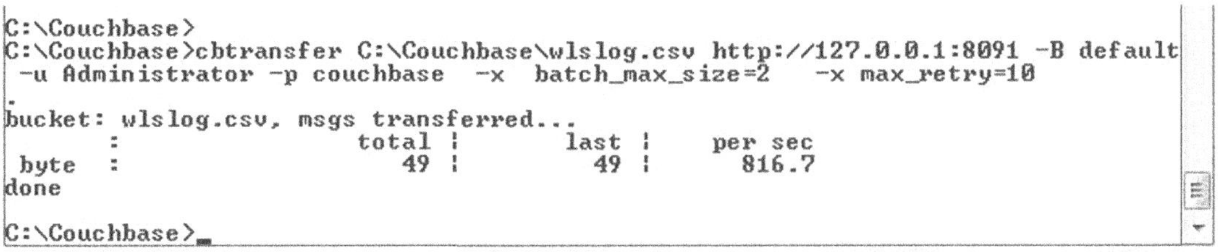

```
C:\Couchbase>
C:\Couchbase>cbtransfer C:\Couchbase\wlslog.csv http://127.0.0.1:8091 -B default
 -u Administrator -p couchbase  -x  batch_max_size=2   -x max_retry=10

bucket: wlslog.csv, msgs transferred...
         :               total :        last :      per sec
 byte    :                  49 :          49 :        816.7
done
C:\Couchbase>_
```

***Figure 11-6.*** *Running the cbtransfer tool command*

# Displaying JSON Data in Couchbase

In the Couchbase Server Admin Console the Item Count should get listed as 7 as shown in Figure 11-7.

***Figure 11-7.*** *The default bucket with Item Count as 7*

Click on the Documents button for the default bucket. The 7 rows of data in Oracle Database are listed as being transferred as 7 documents as shown in Figure 11-8.

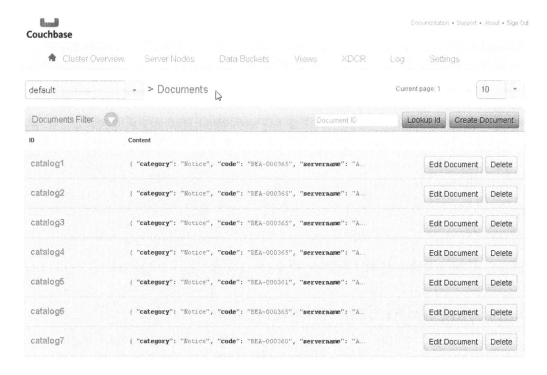

***Figure 11-8.*** *The 7 documents transferred to Couchbase Server*

Click on a catalog1 document. The JSON data in the document gets listed as shown in Figure 11-9.

***Figure 11-9.*** *The catalog1 JSON Document*

# Summary

In this chapter we transferred Oracle Database table data to Couchbase Server. First, we created an Oracle Database table. As a direct data transfer tool is not available, first we exported the Oracle Database table to a CSV file. Subsequently the CSV file data is transferred to Couchbase Server using the Couchbase *cbtransfer* tool. In the next chapter we shall use the Couchbase Hadoop Connector to transfer data between Couchbase and HDFS.

# CHAPTER 12

# Using the Couchbase Hadoop Connector

Apache Sqoop is a tool designed for transferring bulk data between Hadoop and a structured data store. The Couchbase Hadoop Connector is designed for using Sqoop with Couchbase Server. With the Couchbase Hadoop Connector bulk data may be transferred between Hadoop ecosystem-HDFS or Hive, and Couchbase data store. Typically, Hadoop is used for performing data analytics on the data stored in Couchbase Server. In this chapter we shall discuss using the Couchbase Hadoop Connector to transfer data between Couchbase Server and Hadoop HDFS. The chapter covers the following topics.

- Setting Up the Environment
- Installing Couchbase Hadoop Connector
- Listing Tables in Couchbase Server
- Exporting from HDFS to Couchbase
- Importing into HDFS from Couchbase

## Setting Up the Environment

The following software is required for this chapter.

- Couchbase Server
- Couchbase Hadoop Connector
- Sqoop
- Hadoop
- Java 7

Hadoop including Sqoop is designed to be used on Linux. If using Windows, Hadoop may still be used by using Cygwin, a Linux-like environment for using Windows for running software designed for POSIX systems. Download and install Cygwin from `http://cygwin.com/install.html`. When installing Cygwin install the `Net::OpenSSH` package. Cygwin should be installed in a directory without spaces in the directory path. Create a sub-directory without spaces in the `/cygwin64/home` directory. We shall install Couchbase Server, Hadoop, Sqoop, and the Couchbase Hadoop Connector on Oracle Linux 6.5, which is based on Red Hat Linux. Another Linux distribution may also be used. Also, Sqoop 1 is used in this chapter instead of Sqoop 2 as Sqoop 2 currently lacks some of the features of Sqoop 1.

# Installing Couchbase Server on Linux

We have used Couchbase Server 32-bit Red Hat 6 Community Edition 2.2.0 Release in this chapter. If another Linux distribution is used the Couchbase Server download for the Linux distribution should be used.

1.   Install Couchbase Server with the following commands.

     rpm -i couchbase-server-community_2.2.0_x86.rpm

As indicated by the output Couchbase Server gets installed as shown in Figure 12-1.

```
[root@localhost couchbase]# rpm -i couchbase-server-community_2.2.0_x86.rpm
Minimum RAM required   : 4 GB
System RAM configured : 2070564 kB

Minimum number of processors required : 4 cores
Number of processors on the system    : 1 cores

Starting couchbase-server[  OK  ]

You have successfully installed Couchbase Server.
Please browse to http://localhost.oraclelinux:8091/ to configure your server.
Please refer to http://couchbase.com for additional resources.

Please note that you have to update your firewall configuration to
allow connections to the following ports: 11211, 11210, 11209, 4369,
8091, 8092 and from 21100 to 21299.

By using this software you agree to the End User License Agreement.
See /opt/couchbase/LICENSE.txt.

[root@localhost couchbase]# █
```

***Figure 12-1.*** *Installing Couchbase Server on Linux*

2.   Log in to the Couchbase Server Console with URL http://localhost:8091. Click on SETUP as shown in Figure 12-2.

***Figure 12-2.*** *Clicking on SETUP button*

3.  In CONFIGURE SERVER the Database Path is specified. The Hostname is 127.0.0.1 as shown in Figure 12-3.

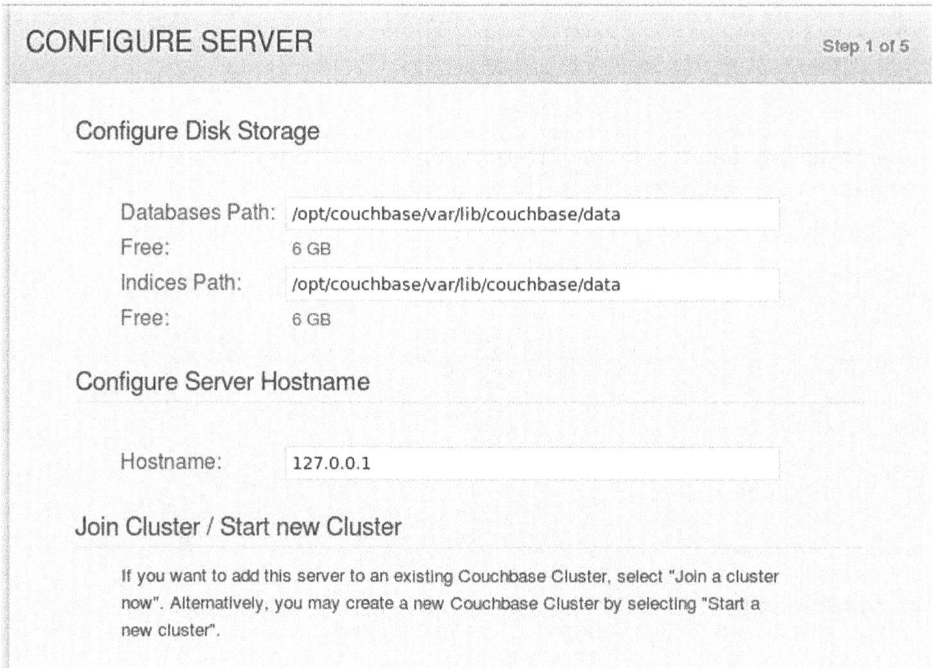

CONFIGURE SERVER                                                        Step 1 of 5

Configure Disk Storage

Databases Path:   /opt/couchbase/var/lib/couchbase/data
Free:             6 GB
Indices Path:     /opt/couchbase/var/lib/couchbase/data
Free:             6 GB

Configure Server Hostname

Hostname:         127.0.0.1

Join Cluster / Start new Cluster

If you want to add this server to an existing Couchbase Cluster, select "Join a cluster now". Alternatively, you may create a new Couchbase Cluster by selecting "Start a new cluster".

*Figure 12-3.* *Configuring Couchbase Server*

4.  In Start new Cluster header, the Start a new cluster is selected as shown in Figure 12-4. Click on Next.

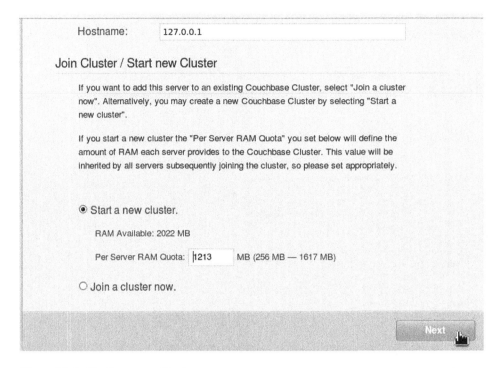

***Figure 12-4.*** *Starting a new Cluster*

5. In SAMPLE BUCKETS the Sample Data and MapReduce samples are listed as shown in Figure 12-5. We won't be using the samples but adding new datasets. Click on Next.

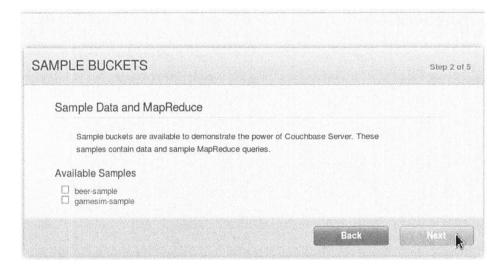

***Figure 12-5.*** *Listing the Sample Buckets*

6. In CREATE DEFAULT BUCKET the Bucket name is set as default and the Bucket Type is Couchbase as shown in Figure 12-6.

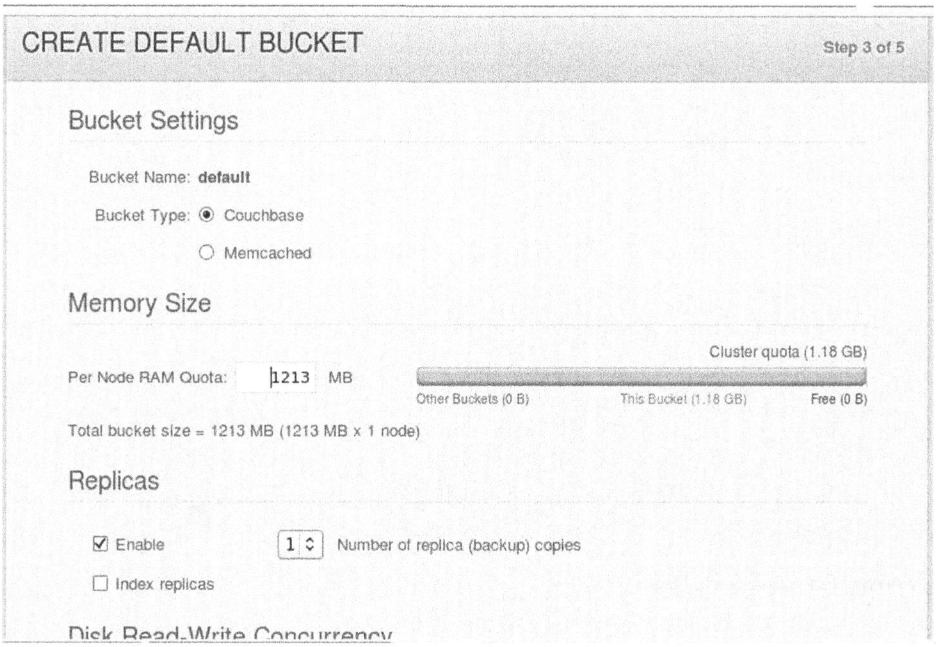

**CREATE DEFAULT BUCKET**                                Step 3 of 5

Bucket Settings

Bucket Name: **default**

Bucket Type: ● Couchbase
              ○ Memcached

Memory Size

Cluster quota (1.18 GB)

Per Node RAM Quota:  1213  MB

Other Buckets (0 B)      This Bucket (1.18 GB)      Free (0 B)

Total bucket size = 1213 MB (1213 MB x 1 node)

Replicas

☑ Enable      1 ⇕  Number of replica (backup) copies

☐ Index replicas

Disk Read-Write Concurrency

*Figure 12-6.* *Selecting Data Bucket Name and Type*

7. The Per Node RAM Quota is also set. The Number of replica copies is set as 1. Click on Next as shown in Figure 12-7.

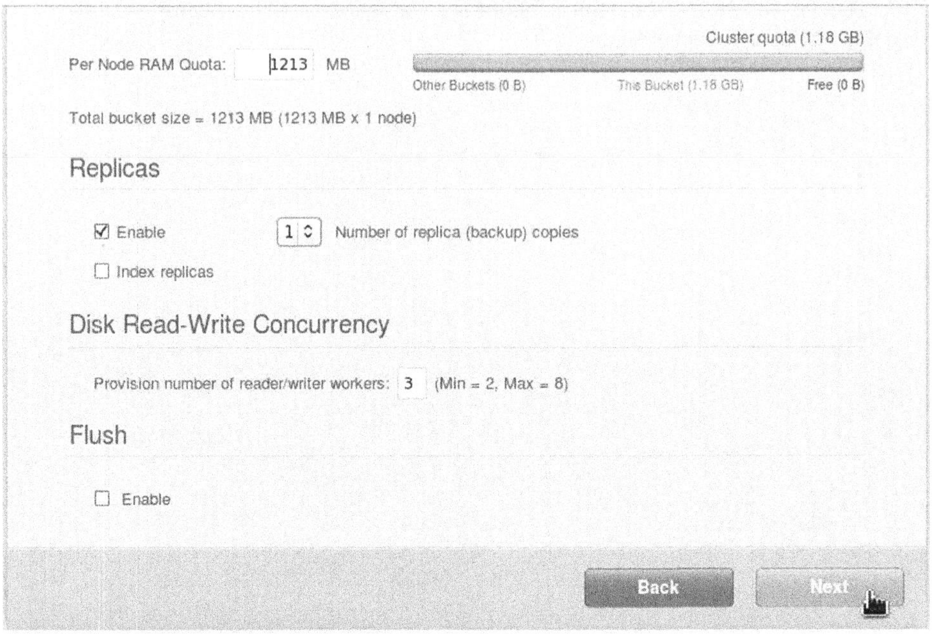

***Figure 12-7.*** *Specifying Per Node RAM Quota*

8. In Update Notifications click on Next as shown in Figure 12-8.

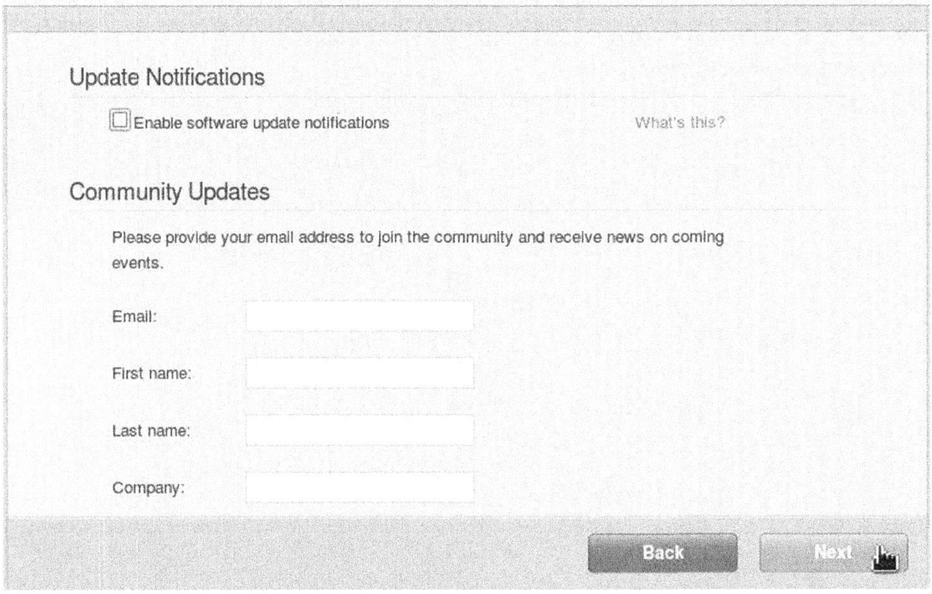

***Figure 12-8.*** *Update Notifications*

9. Specify Username (Administrator) and Password and click on Next as shown in Figure 12-9.

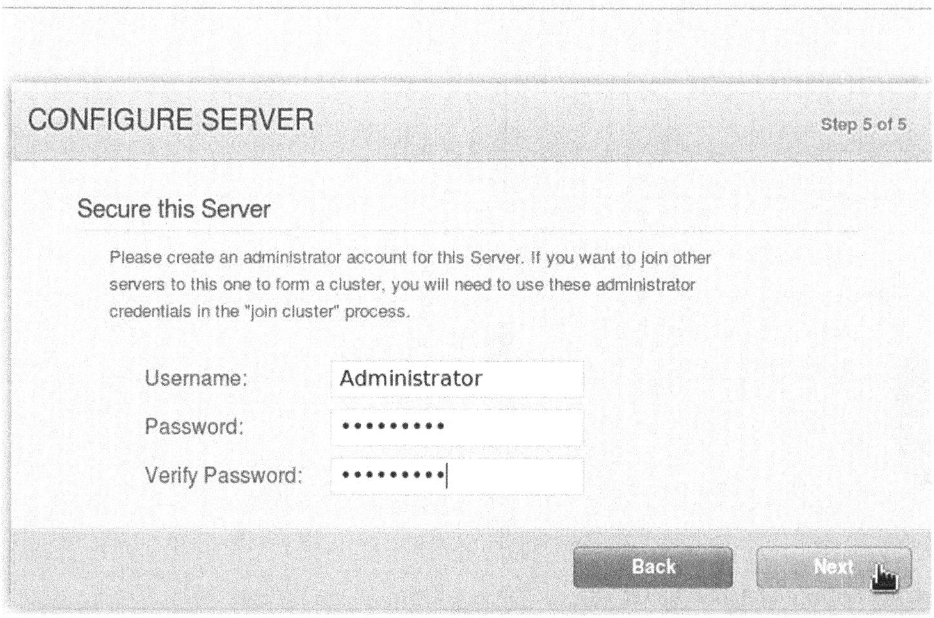

*Figure 12-9.* *Specifying Username and Password*

In Cluster Overview the RAM (Total Allocated, In Use, Unused, and Unallocated) and Disk (Unused Free Space, In Use, Other Data and Free) are listed as shown in Figure 12-10.

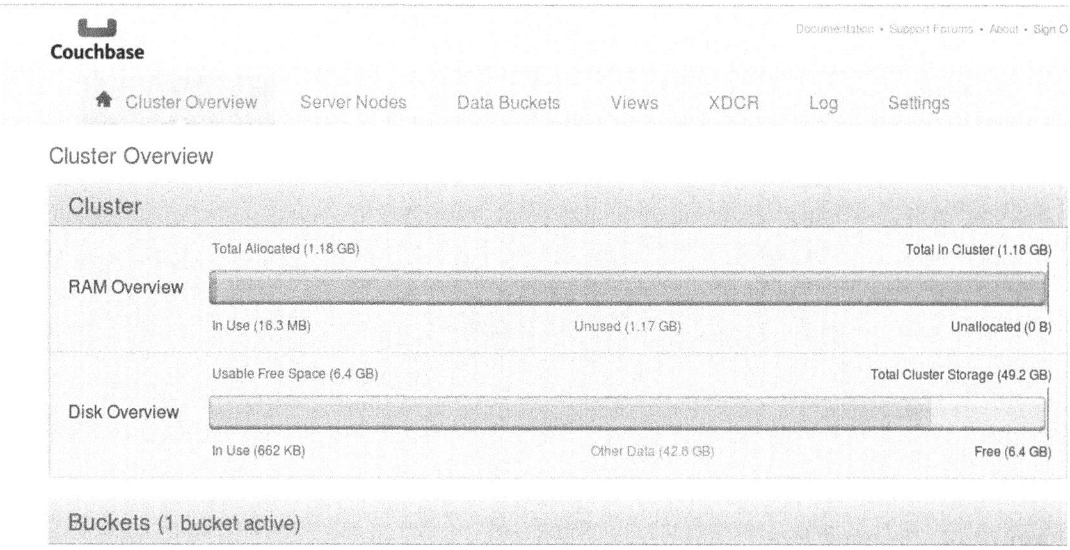

*Figure 12-10.* *Cluster Overview*

On the same page one active server and one active bucket are listed as shown in Figure 12-11.

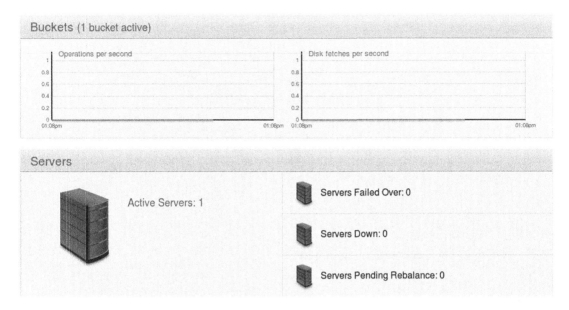

*Figure 12-11.* *Active Server and Active Bucket*

In Data Buckets/Couchbase Buckets, the default bucket is listed along with the number of Nodes, Item Count, and other stats about the bucket as shown in Figure 12-12.

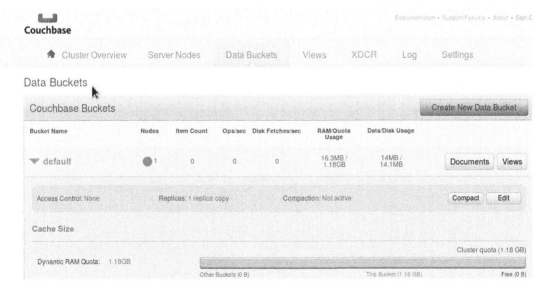

*Figure 12-12.* *The default bucket*

# Installing Hadoop and Sqoop

First, we need to install the Hadoop ecosystem in which to install the Couchbase Hadoop Connector. We shall use CDH 4.6 and download Hadoop 2.0.0 CDH 4.6 and Sqoop 1.4.3 CDH 4.6 TAR files from http://archive.cloudera.com/cdh4/cdh/4/.

1. Create a directory /couchbase for installing Couchbase Hadoop Connector and set its permissions.

```
mkdir /couchbase
chmod -R 777 /couchbase
cd /couchbase
```

2. Install Java 7 by downloading the Java 7 Linux gz file and subsequently running the following tar command.

```
tar zxvf jdk-7u55-linux-i586.gz
```

3. Download and install Hadoop 2.0.0 CDH 4.6 with the following commands.

```
wget http://archive.cloudera.com/cdh4/cdh/4/hadoop-2.0.0-cdh4.6.0.tar.gz
tar -xvf hadoop-2.0.0-cdh4.6.0.tar.gz
```

4. Create Symlinks for the Hadoop MapReduce 2 (MRv2) bin and conf directories.

```
ln -s /couchbase/hadoop-2.0.0-cdh4.6.0/bin /couchbase/hadoop-2.0.0-cdh4.6.0/
share/hadoop/mapreduce2/bin
ln -s /couchbase/hadoop-2.0.0-cdh4.6.0/etc/hadoop /couchbase/hadoop-2.0.0-
cdh4.6.0/share/hadoop/mapreduce2/conf
```

5. Download and install Sqoop 1.4.3 CDH 4.6 with the following commands.

```
wget http://archive-primary.cloudera.com/cdh4/cdh/4/sqoop-1.4.3-cdh4.6.0.tar.gz
tar -xzf sqoop-1.4.3-cdh4.6.0.tar.gz
```

6. Set the environment variables for Java, Hadoop, and Sqoop in the bash shell.

```
vi ~/.bashrc
export HADOOP_HOME=/couchbase/hadoop-2.0.0-cdh4.6.0/share/hadoop/mapreduce2
export HADOOP_PREFIX=/couchbase/hadoop-2.0.0-cdh4.6.0
export HADOOP_CONF=$HADOOP_PREFIX/etc/hadoop
export SQOOP_HOME=/couchbase/sqoop-1.4.3-cdh4.6.0
export JAVA_HOME=/couchbase/jdk1.7.0_55
export HADOOP_MAPRED_HOME=/couchbase/hadoop-2.0.0-cdh4.6.0/bin
export HADOOP_CLASSPATH=$HADOOP_HOME/*:$HADOOP_HOME/lib/*
export PATH=$PATH:$HADOOP_HOME/bin:$HADOOP_MAPRED_HOME:$SQOOP_HOME/bin
```

7. Copy the required Hadoop Jar files to the Sqoop classpath.

```
cp /couchbase/hadoop-2.0.0-cdh4.6.0/share/hadoop/mapreduce2/*
/couchbase/sqoop-1.4.3-cdh4.6.0/lib
cp /couchbase/hadoop-2.0.0-cdh4.6.0/share/hadoop/mapreduce1/lib/
hadoop-common-2.0.0-cdh4.6.0.jar  /couchbase/sqoop-1.4.3-cdh4.6.0/lib
cp /couchbase/hadoop-2.0.0-cdh4.6.0/share/hadoop/mapreduce1/lib/
commons-configuration-1.6.jar  /couchbase/sqoop-1.4.3-cdh4.6.0/lib
cp /couchbase/hadoop-2.0.0-cdh4.6.0/share/hadoop/mapreduce1/lib/
hadoop-auth-2.0.0-cdh4.6.0.jar  /couchbase/sqoop-1.4.3-cdh4.6.0/lib
cp /couchbase/hadoop-2.0.0-cdh4.6.0/share/hadoop/mapreduce1/lib/
slf4j-api-1.6.1.jar  /couchbase/sqoop-1.4.3-cdh4.6.0/lib
cp /couchbase/hadoop-2.0.0-cdh4.6.0/share/hadoop/mapreduce1/lib/
commons-httpclient-3.1.jar  /couchbase/sqoop-1.4.3-cdh4.6.0/lib
cp /couchbase/hadoop-2.0.0-cdh4.6.0/share/hadoop/mapreduce1/lib/
commons-collections-3.2.1.jar  /couchbase/sqoop-1.4.3-cdh4.6.0/lib
```

8. The sqoop tool usage may be listed with the following command.

```
sqoop help
```

The output from the command lists the available commands as shown in Figure 12-13.

```
14/06/20 17:54:40 INFO sqoop.Sqoop: Running Sqoop version: 1.4.3-cdh4.6.0
usage: sqoop COMMAND [ARGS]

Available commands:
  codegen            Generate code to interact with database records
  create-hive-table  Import a table definition into Hive
  eval               Evaluate a SQL statement and display the results
  export             Export an HDFS directory to a database table
  help               List available commands
  import             Import a table from a database to HDFS
  import-all-tables  Import tables from a database to HDFS
  job                Work with saved jobs
  list-databases     List available databases on a server
  list-tables        List available tables in a database
  merge              Merge results of incremental imports
  metastore          Run a standalone Sqoop metastore
  version            Display version information

See 'sqoop help COMMAND' for information on a specific command.
[root@localhost couchbase]# █
```

*Figure 12-13.* *Sqoop Commands*

In subsequent sections we shall discuss the Sqoop commands listed in Table 12-1.

*Table 12-1.* *Sqoop Commands*

| Sqoop command | Description |
| --- | --- |
| list-tables | Lists the tables in the Couchbase Server. |
| export | Exports Key-value pairs from HDFS to Couchbase Server. |
| import | Imports Key-Value pairs from Couchbase Server to HDFS. |

# Installing Couchbase Hadoop Connector

The procedure to install Couchbase Hadoop Connector is as follows.

1.  Download the CDH4 compatible Couchbase Hadoop Connector couchbase-hadoop-plugin-1.1-dp3.zip file from `http://packages.couchbase.com.s3.amazonaws.com/clients/connectors/couchbase-hadoop-plugin-1.1-dp3.zip`.

2.  Extract the couchbase-hadoop-plugin-1.1-dp3.zip file to the `/couchbase` directory.

    ```
    cd /couchbase
    unzip couchbase-hadoop-plugin-1.1-dp3.zip
    ```

3.  Copy the Couchbase Hadoop Connector Jar files from the `/couchbase` directory to the Sqoop classpath, which is the `/couchbase/sqoop-1.4.3-cdh4.6.0/lib` directory.

    ```
    cp /couchbase/*.jar  /couchbase/sqoop-1.4.3-cdh4.6.0/lib
    ```

4.  Copy the `couchbase-config.xml` from the Couchbase Hadoop Connector installation to the `/couchbase/sqoop-1.4.3-cdh4.6.0/conf` directory.

    ```
    cp /couchbase/couchbase-config.xml  /couchbase/sqoop-1.4.3-cdh4.6.0/conf
    ```

5.  Create a directory `/couchbase/sqoop-1.4.3-cdh4.6.0/conf/managers.d` and copy the `couchbase-manager.xml` from the `/couchbase` directory to the `managers.d` sub-directory.

    ```
    mkdir /couchbase/sqoop-1.4.3-cdh4.6.0/conf/managers.d
    cp /couchbase/couchbase-manager.xml  /couchbase/sqoop-1.4.3-cdh4.6.0/
    conf/managers.d
    ```

6.  Copying the Couchbase Hadoop Connector jars and configuration files to the Sqoop installation completes the installation of Couchbase Hadoop Connector. To verify the Couchbase Hadoop Connector installation, run the following command.

    ```
    sh ./install.sh %SQOOP_HOME%
    ```

    The output from the command indicates that the required files are found in Sqoop and the connector installation is successful as shown in Figure 12-14.

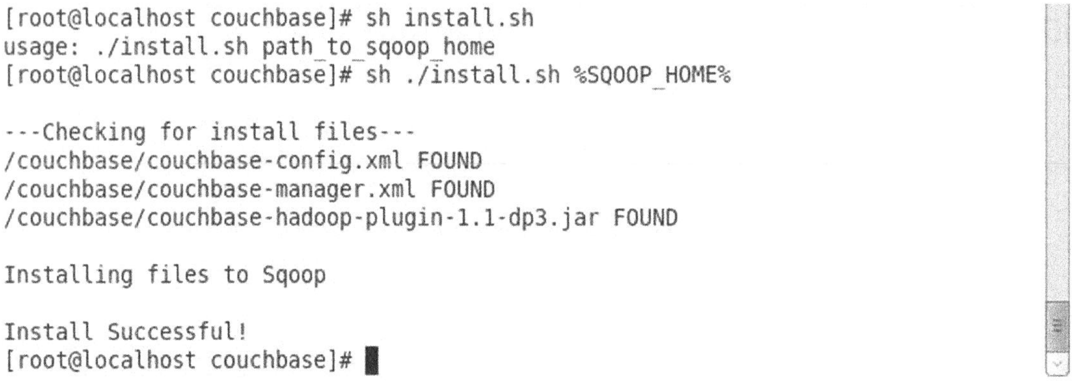

```
[root@localhost couchbase]# sh install.sh
usage: ./install.sh path_to_sqoop_home
[root@localhost couchbase]# sh ./install.sh %SQOOP_HOME%

---Checking for install files---
/couchbase/couchbase-config.xml FOUND
/couchbase/couchbase-manager.xml FOUND
/couchbase/couchbase-hadoop-plugin-1.1-dp3.jar FOUND

Installing files to Sqoop

Install Successful!
[root@localhost couchbase]# █
```

*Figure 12-14.* *Installing Couchbase Hadoop Connector*

If the couchbase-manager.xml file is not found in the sqoop-1.4.3-cdh4.6.0\conf\managers.d directory, the following error message is generated.

```
ERROR tool.BaseSqoopTool: Got error creating database manager: java.io.IOException: No
manager for connect string: http://localhost:8091/pools
        at org.apache.sqoop.ConnFactory.getManager(ConnFactory.java:185)
```

# Listing Tables in Couchbase Server

Sqoop is designed for relational databases in which tables are the norm, which as we know is not how Couchbase is structured. The Couchbase Hadoop Connector accepts the –table option and uses it for imports only as the tap stream to import from. For exports any value for the –table option may be specified as it is ignored by the connector. For the sqoop import command the following values listed in Table 12-2 may be specified for the –table option.

*Table 12-2.* *Sqoop import -table Options*

| Table | Description |
|---|---|
| DUMP | Contains all key-value pairs in the Couchbase Server and the sqoop import command imports all the key-value pairs into HDFS. |
| BACKFILL_nn | Contains all key mutations for a specified time (nn) in minutes. Sqoop import command with the BACKFILL_nn streams all subsequent key mutations into HDFS for the specified time. For example, BACKFILL_5 as –table option value streams key mutations for the subsequent 5 minutes. |

Though Couchbase Server does not store data in tables the DUMP and BACKFILL_nn may be used as values to the –table option. To list the tables in the Couchbase Server run the following command; the –connect option specifies the connection URL to the Couchbase Server and should include the Ipv4 address (10.0.2.15 in the example) instead of "localhost" if Linux is running in a virtual machine such as Oracle VirtualBox.

```
sqoop list-tables --connect http://10.0.2.15:8091/pools
```

The DUMP and BACKFILL_NN tables get listed as shown in Figure 12-15.

```
[root@localhost couchbase]# sqoop list-tables --connect http://10.0.2.15:8091/po
ols
Warning: /usr/lib/hbase does not exist! HBase imports will fail.
Please set $HBASE_HOME to the root of your HBase installation.
Warning: /usr/lib/hcatalog does not exist! HCatalog jobs will fail.
Please set $HCAT_HOME to the root of your HCatalog installation.
Warning: /couchbase/sqoop-1.4.3-cdh4.6.0/../accumulo does not exist! Accumulo im
ports will fail.
Please set $ACCUMULO_HOME to the root of your Accumulo installation.
Warning: /couchbase/sqoop-1.4.3-cdh4.6.0/../zookeeper does not exist! Accumulo i
mports will fail.
Please set $ZOOKEEPER_HOME to the root of your Zookeeper installation.
14/06/20 08:55:54 INFO sqoop.Sqoop: Running Sqoop version: 1.4.3-cdh4.6.0
DUMP
BACKFILL_NN
[root@localhost couchbase]# 
```

***Figure 12-15.*** *Listing Tables in Couchbase*

Next, we shall export key-value pairs from HDFS to Couchbase Server.

# Exporting from HDFS to Couchbase

To export data from HDFS to Couchbase Server the data has to be available in HDFS in a key-value format. Store data to be exported in a file (catalog.json) on the local filesystem.

```
"journal": "Oracle Magazine",
"publisher": "Oracle Publishing",
"edition": "November-December 2013",
"title": "Engineering as a Service",
"author": "David A. Kelly",
```

In JSON format the fields are delimited by the ':' character and the records are demarcated by the ',' character. A field value must be enclosed in double quotes (""), though a key may be un-enclosed or enclosed in single quotes. Create a directory on the HDFS to store the catalog.json using the following command.

```
hdfs dfs -mkdir /catalog
```

Put or copy the catalog.json from the local filesystem to the HDFS using the following command.

```
hdfs dfs -put  catalog.json /catalog/catalog.json
```

The data in catalog.json in HDFS may be listed with the following command.

```
hdfs dfs -cat /catalog/catalog.json
```

The output from the command lists the key-value pairs in the catalog.json as shown in Figure 12-16.

```
[root@localhost couchbase]# hdfs dfs -mkdir /catalog
14/06/20 09:00:47 WARN util.NativeCodeLoader: Unable to load native-hadoop libra
ry for your platform... using builtin-java classes where applicable
[root@localhost couchbase]# hdfs dfs -put  catalog.json /catalog
14/06/20 09:01:21 WARN util.NativeCodeLoader: Unable to load native-hadoop libra
ry for your platform... using builtin-java classes where applicable
[root@localhost couchbase]# hdfs dfs -cat /catalog/catalog.json
14/06/20 09:01:40 WARN util.NativeCodeLoader: Unable to load native-hadoop libra
ry for your platform... using builtin-java classes where applicable
"journal": "Oracle Magazine",
"publisher": "Oracle Publishing",
"edition": "November-December 2013",
"title": "Engineering as a Service",
"author": "David A. Kelly",
[root@localhost couchbase]# █
```

***Figure 12-16.*** *Listing the Key-Value Pairs in catalog.json*

To export the key-value pairs in catalog.json to the Couchbase Server run the following sqoop export command. The –connect option specifies the connection URL to Couchbase Server. The –export-dir option specifies the directory in HDFS that is to be exported to Couchbase Server. The –table option is required to be specified though the value specified is not used, therefore any arbitrary value may be specified. The --fields-terminated-by option must be set to ':' and the --lines-terminated-by option to ','.

```
sqoop export --connect http://10.0.2.15:8091/pools   --export-dir /catalog   --table catalog
--fields-terminated-by : --lines-terminated-by ,
```

The output from the command indicates that five records have been exported to Couchbase Server as shown in Figure 12-17.

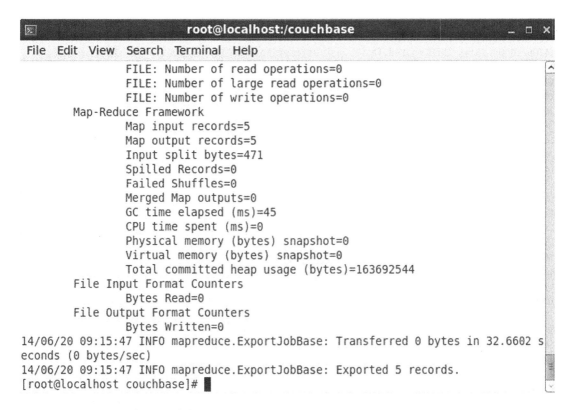

```
                  FILE: Number of read operations=0
                  FILE: Number of large read operations=0
                  FILE: Number of write operations=0
          Map-Reduce Framework
                  Map input records=5
                  Map output records=5
                  Input split bytes=471
                  Spilled Records=0
                  Failed Shuffles=0
                  Merged Map outputs=0
                  GC time elapsed (ms)=45
                  CPU time spent (ms)=0
                  Physical memory (bytes) snapshot=0
                  Virtual memory (bytes) snapshot=0
                  Total committed heap usage (bytes)=163692544
          File Input Format Counters
                  Bytes Read=0
          File Output Format Counters
                  Bytes Written=0
14/06/20 09:15:47 INFO mapreduce.ExportJobBase: Transferred 0 bytes in 32.6602 s
econds (0 bytes/sec)
14/06/20 09:15:47 INFO mapreduce.ExportJobBase: Exported 5 records.
[root@localhost couchbase]#
```

*Figure 12-17.*  *Listing the Records exported to Couchbase Server*

A more detailed output from the export command is listed:

```
14/06/20 09:15:37 INFO mapreduce.Job: The url to track the job: http://localhost:8080/
14/06/20 09:15:37 INFO mapreduce.Job: Running job: job_local193560694_0001
14/06/20 09:15:38 INFO mapred.LocalJobRunner: OutputCommitter set in config null
14/06/20 09:15:38 INFO mapred.LocalJobRunner: OutputCommitter is org.apache.hadoop
.mapreduce.lib.output.FileOutputCommitter
14/06/20 09:15:39 INFO mapreduce.Job: Job job_local193560694_0001 running in uber mode : false
14/06/20 09:15:39 INFO mapred.LocalJobRunner: Waiting for map tasks
14/06/20 09:15:39 INFO mapreduce.Job:   map 0% reduce 0%
14/06/20 09:15:39 INFO mapred.LocalJobRunner: Starting task: attempt_local193560694_0001
_m_000000_0
14/06/20 09:15:40 INFO mapred.Task:  Using ResourceCalculatorProcessTree : [ ]
14/06/20 09:15:40 INFO mapred.MapTask: Processing split: Paths:/catalog/catalog.json:126+22,
/catalog/catalog.json:148+23
14/06/20 09:15:44 INFO client.CouchbaseConnection: Added {QA sa=/10.0.2.15:11210, #Rops=0,
#Wops=0, #iq=0, topRop=null, topWop=null, toWrite=0, interested=0} to connect queue
14/06/20 09:15:44 INFO client.CouchbaseConnection: Connection state changed for
sun.nio.ch.SelectionKeyImpl@1c0280b
14/06/20 09:15:45 INFO mapreduce.AutoProgressMapper: Auto-progress thread is
finished. keepGoing=false
```

```
14/06/20 09:15:45 INFO mapred.LocalJobRunner:
14/06/20 09:15:45 INFO client.CouchbaseConnection: Shut down Couchbase client
14/06/20 09:15:45 INFO mapred.Task: Task:attempt_local193560694_0001_m_000000_0 is done.
And is in the process of committing
14/06/20 09:15:45 INFO mapred.LocalJobRunner: map
14/06/20 09:15:45 INFO mapred.Task: Task 'attempt_local193560694_0001_m_000000_0' done.
14/06/20 09:15:45 INFO mapred.LocalJobRunner: Finishing task: attempt_local193560694_0001
_m_000000_0
14/06/20 09:15:45 INFO mapred.LocalJobRunner: Starting task: attempt_local193560694_0001
_m_000001_0
14/06/20 09:15:45 INFO mapred.Task:  Using ResourceCalculatorProcessTree : [ ]
14/06/20 09:15:45 INFO mapred.MapTask: Processing split: Paths:/catalog/catalog.json:0+42
14/06/20 09:15:46 INFO client.CouchbaseConnection: Added {QA sa=/10.0.2.15:11210, #Rops=0,
#Wops=0, #iq=0, topRop=null, topWop=null, toWrite=0, interested=0} to connect queue
14/06/20 09:15:46 INFO client.CouchbaseConnection: Connection state changed for sun.nio.ch
.SelectionKeyImpl@e8c6f
14/06/20 09:15:46 INFO mapreduce.AutoProgressMapper: Auto-progress thread is finished.
keepGoing=false
14/06/20 09:15:46 INFO mapred.LocalJobRunner:
14/06/20 09:15:46 INFO client.CouchbaseConnection: Shut down Couchbase client
14/06/20 09:15:46 INFO mapred.Task: Task:attempt_local193560694_0001_m_000001_0 is done.
And is in the process of committing
14/06/20 09:15:46 INFO mapred.LocalJobRunner: map
14/06/20 09:15:46 INFO mapred.Task: Task 'attempt_local193560694_0001_m_000001_0' done.
14/06/20 09:15:46 INFO mapred.LocalJobRunner: Finishing task: attempt_local193560694_0001
_m_000001_0
14/06/20 09:15:46 INFO mapred.LocalJobRunner: Starting task: attempt_local193560694_0001
_m_000002_0
14/06/20 09:15:46 INFO mapreduce.Job:  map 100% reduce 0%
14/06/20 09:15:46 INFO mapred.Task:  Using ResourceCalculatorProcessTree : [ ]
14/06/20 09:15:46 INFO mapred.MapTask: Processing split: Paths:/catalog/catalog.json:42+42
14/06/20 09:15:46 INFO client.CouchbaseConnection: Added {QA sa=/10.0.2.15:11210, #Rops=0,
#Wops=0, #iq=0, topRop=null, topWop=null, toWrite=0, interested=0} to connect queue
14/06/20 09:15:46 INFO client.CouchbaseConnection: Connection state changed for sun.nio
.ch.SelectionKeyImpl@5c414e
14/06/20 09:15:46 INFO mapreduce.AutoProgressMapper: Auto-progress thread is
finished. keepGoing=false
14/06/20 09:15:46 INFO mapred.LocalJobRunner:
14/06/20 09:15:46 INFO client.CouchbaseConnection: Shut down Couchbase client
14/06/20 09:15:46 INFO mapred.Task: Task:attempt_local193560694_0001_m_000002_0 is done.
And is in the process of committing
14/06/20 09:15:46 INFO mapred.LocalJobRunner: map
14/06/20 09:15:46 INFO mapred.Task: Task 'attempt_local193560694_0001_m_000002_0' done.
14/06/20 09:15:46 INFO mapred.LocalJobRunner: Finishing task: attempt_local193560694_0001
_m_000002_0
14/06/20 09:15:46 INFO mapred.LocalJobRunner: Starting task: attempt_local193560694_0001
_m_000003_0
14/06/20 09:15:46 INFO mapred.Task:  Using ResourceCalculatorProcessTree : [ ]
14/06/20 09:15:46 INFO mapred.MapTask: Processing split: Paths:/catalog/catalog.json:84+42
14/06/20 09:15:46 INFO client.CouchbaseConnection: Added {QA sa=/10.0.2.15:11210, #Rops=0,
#Wops=0, #iq=0, topRop=null, topWop=null, toWrite=0, interested=0} to connect queue
```

```
14/06/20 09:15:46 INFO client.CouchbaseConnection: Connection state changed for
sun.nio.ch.SelectionKeyImpl@dfc71d
14/06/20 09:15:46 INFO mapreduce.AutoProgressMapper: Auto-progress thread is
finished. keepGoing=false
14/06/20 09:15:46 INFO mapred.LocalJobRunner:
14/06/20 09:15:46 INFO client.CouchbaseConnection: Shut down Couchbase client
14/06/20 09:15:46 INFO mapred.Task: Task:attempt_local193560694_0001_m_000003_0 is done.
And is in the process of committing
14/06/20 09:15:46 INFO mapred.LocalJobRunner: map
14/06/20 09:15:46 INFO mapred.Task: Task 'attempt_local193560694_0001_m_000003_0' done.
14/06/20 09:15:46 INFO mapred.LocalJobRunner: Finishing task: attempt_local193560694_0001
_m_000003_0
14/06/20 09:15:46 INFO mapred.LocalJobRunner: Map task executor complete.
14/06/20 09:15:47 INFO mapreduce.Job: Job job_local193560694_0001 completed successfully
14/06/20 09:15:47 INFO mapreduce.Job: Counters: 18
    File System Counters
        FILE: Number of bytes read=77331208
        FILE: Number of bytes written=78780032
        FILE: Number of read operations=0
        FILE: Number of large read operations=0
        FILE: Number of write operations=0
    Map-Reduce Framework
        Map input records=5
        Map output records=5
        Input split bytes=471
        Spilled Records=0
        Failed Shuffles=0
        Merged Map outputs=0
        GC time elapsed (ms)=45
        CPU time spent (ms)=0
        Physical memory (bytes) snapshot=0
        Virtual memory (bytes) snapshot=0
        Total committed heap usage (bytes)=163692544
    File Input Format Counters
        Bytes Read=0
    File Output Format Counters
        Bytes Written=0
14/06/20 09:15:47 INFO mapreduce.ExportJobBase: Transferred 0 bytes in 32.6602 seconds
(0 bytes/sec)
14/06/20 09:15:47 INFO mapreduce.ExportJobBase: Exported 5 records.
```

The following error message if generated during export may be ignored as indicated by the error message the "issue might not necessarily be caused by current input" and is "due to the batching nature of export."

```
INFO mapreduce.ExportJobBase: Exported 4 records.
14/01/01 17:49:56 ERROR tool.ExportTool: Error during export: Export job failed!
ERROR mapreduce.TextExportMapper: This issue might not necessarily be caused by current input
14/01/01 17:49:56 ERROR mapreduce.TextExportMapper: due to the batching nature of export.
```

Log in to the Couchbase Administration Console and select the Data Buckets link. The "default" bucket has Item Count as 5 as shown in Figure 12-18.

***Figure 12-18.*** *Displaying the Item Count in the default Bucket*

Click on the Documents button for the "default" button to list the five key-value records exported to Couchbase Server as shown in Figure 12-19.

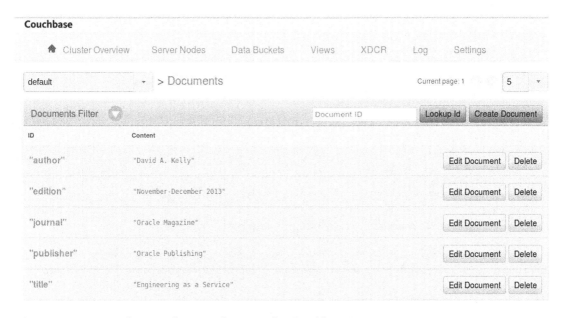

***Figure 12-19.*** *Listing the Key-Value Records Exported to Couchbase Server*

A key may be un-enclosed (or enclosed in single quotes), but the field value must be enclosed in double quotes in the key-value records exported from HDFS. To demonstrate copy the following listing to catalog2.json file.

```
journal:"Oracle Magazine",
publisher:"Oracle Publishing",
edition:"November December 2013",
title:"Engineering as a Service",
author:"David A. Kelly",
journal2: "Oracle Magazine",
publisher2: "Oracle Publishing",
edition2: "November-December 2013",
title2: "Quintessential and Collaborative",
author2: "Tom Haunert",
```

A key must be unique, but the key value may be the same as some other field value. Remove the catalog.json file from HDFS and put the catalog2.json file into HDFS.

```
hdfs dfs -rm /catalog/catalog.json
hdfs dfs -put  catalog2.json /catalog
```

Run the Sqoop command to export from HDFS to Couchbase Server.

```
sqoop export --connect http://10.0.2.15:8091/pools    --export-dir /catalog    --table catalog
--fields-terminated-by : --lines-terminated-by ,
```

The key/value pairs get exported to Couchbase Server as shown in Figure 12-20.

*Figure 12-20.* *Listing the key/value Pairs exported to Couchbase*

The key-value pairs exported from HDFS must be in the specified format in --fields-terminated-by and --lines-terminated-by options. If some other format is used the fields may not get parsed and the following exception might get generated.

```
java.util.NoSuchElementException
        at java.util.ArrayList$Itr.next(ArrayList.java:794)
        at catalog.__loadFromFields(catalog.java:198)
        at catalog.parse(catalog.java:147)
```

# Importing into HDFS from Couchbase

In this section we import key-value pairs from Couchbase Server to HDFS. The key-value pairs in Couchbase Server are not directly importable, but may be imported via the DUMP and BACKFILL_nn tables, which were discussed earlier.

## Importing the Key-Value Pairs Previously Exported

In this section we shall import back the key-value pairs previously exported from HDFS file catalog.json to Couchbase Server. To import the key-value pairs import the DUMP table using the sqoop import command in which the table is specified with the –table option. The HDFS directory to import into is specified with the --target-dir option as /dump_dir, The --as-textfile indicates that the key-value pairs are to be imported as text file. The other options are --as-avrodatafile and --as-sequencefile. Data may be imported in append mode with the –append option. Run the following command to import the DUMP table.

```
sqoop import -Dmapreduce.job.max.split.locations=2048 --connect http://10.0.2.15:8091/pools
--table DUMP --target-dir hdfs://localhost:8020/dump_dir --as-textfile
```

The mapreduce.job.max.split.locations property has been set to a value higher than the default value of 10. If the default value is used the following exception gets generated (some intermediate values have been omitted).

```
ERROR tool.ImportTool: Encountered IOException running import job: java.io.IOException: Max
block location exceeded for split: 0 1 2 3 4 5 6 7 8 9 10          1018 1019 1020 1021 1022
1023  splitsize: 1024 maxsize: 10
```

The value to set mapreduce.job.max.split.locations may be obtained from running the command with the default value and finding the splitsize; set the value to higher than the splitsize. The output from the sqoop import command indicates that five records have been imported as shown in Figure 12-21.

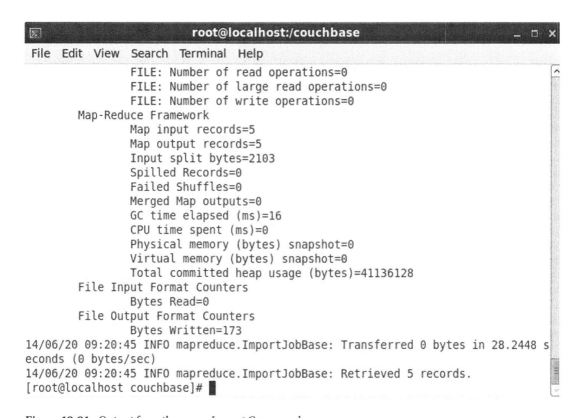

```
                      FILE: Number of read operations=0
                      FILE: Number of large read operations=0
                      FILE: Number of write operations=0
              Map-Reduce Framework
                      Map input records=5
                      Map output records=5
                      Input split bytes=2103
                      Spilled Records=0
                      Failed Shuffles=0
                      Merged Map outputs=0
                      GC time elapsed (ms)=16
                      CPU time spent (ms)=0
                      Physical memory (bytes) snapshot=0
                      Virtual memory (bytes) snapshot=0
                      Total committed heap usage (bytes)=41136128
              File Input Format Counters
                      Bytes Read=0
              File Output Format Counters
                      Bytes Written=173
14/06/20 09:20:45 INFO mapreduce.ImportJobBase: Transferred 0 bytes in 28.2448 s
econds (0 bytes/sec)
14/06/20 09:20:45 INFO mapreduce.ImportJobBase: Retrieved 5 records.
[root@localhost couchbase]# 
```

***Figure 12-21.*** *Output from the sqoop Import Command*

The more detailed output from the command is listed below.

```
14/06/20 09:20:42 INFO mapreduce.Job: The url to track the job: http://localhost:8080/
14/06/20 09:20:42 INFO mapreduce.Job: Running job: job_local352714375_0001
14/06/20 09:20:42 INFO mapred.LocalJobRunner: OutputCommitter set in config null
14/06/20 09:20:43 INFO mapred.LocalJobRunner: OutputCommitter is org.apache.hadoop.mapreduce
.lib.output.FileOutputCommitter
14/06/20 09:20:43 INFO mapred.LocalJobRunner: Waiting for map tasks
14/06/20 09:20:43 INFO mapred.LocalJobRunner: Starting task: attempt_local352714375_0001
_m_000000_0
14/06/20 09:20:43 INFO mapred.Task:  Using ResourceCalculatorProcessTree : [ ]
14/06/20 09:20:43 WARN conf.Configuration: mapreduce.map.class is deprecated. Instead, use
mapreduce.job.map.class
14/06/20 09:20:43 INFO mapreduce.Job: Job job_local352714375_0001 running in uber mode : false
14/06/20 09:20:43 INFO mapreduce.Job:  map 0% reduce 0%
14/06/20 09:20:43 INFO mapred.MapTask: Processing split: 0 1 2 3 4 5 6 7 8 9 10 11 12 13 14
15 16 17 18 19 20 21 ... 1006 1007 1008 1009 1010 1011 1012 1013 1014 1015 1016 1017 1018
1019 1020 1021 1022 1023
14/06/20 09:20:44 INFO client.CouchbaseConnection: Added {QA sa=/10.0.2.15:11210, #Rops=0,
#Wops=0, #iq=0, topRop=null, topWop=null, toWrite=0, interested=0} to connect queue
14/06/20 09:20:44 INFO client.CouchbaseConnection: Connection state changed for
sun.nio.ch.SelectionKeyImpl@d3ec2f
```

```
14/06/20 09:20:45 INFO client.CouchbaseConnection: Shut down Couchbase client
14/06/20 09:20:45 INFO db.CouchbaseRecordReader: All TAP messages have been received.


14/06/20 09:20:45 INFO mapreduce.AutoProgressMapper: Auto-progress thread is finished.
keepGoing=false
14/06/20 09:20:45 INFO mapred.LocalJobRunner:
14/06/20 09:20:45 INFO mapred.Task: Task:attempt_local352714375_0001_m_000000_0 is done.
And is in the process of committing
14/06/20 09:20:45 INFO mapred.LocalJobRunner:
14/06/20 09:20:45 INFO mapred.Task: Task attempt_local352714375_0001_m_000000_0 is allowed
to commit now
14/06/20 09:20:45 INFO output.FileOutputCommitter: Saved output of task
'attempt_local352714375_0001_m_000000_0' to file:/dump_dir/_temporary/0/task_
local352714375_0001_m_000000
14/06/20 09:20:45 INFO mapred.LocalJobRunner: map
14/06/20 09:20:45 INFO mapred.Task: Task 'attempt_local352714375_0001_m_000000_0' done.
14/06/20 09:20:45 INFO mapred.LocalJobRunner: Finishing task: attempt_local352714375_0001
_m_000000_0
14/06/20 09:20:45 INFO mapred.LocalJobRunner: Map task executor complete.
14/06/20 09:20:45 INFO mapreduce.Job:  map 100% reduce 0%
14/06/20 09:20:45 INFO mapreduce.Job: Job job_local352714375_0001 completed successfully
14/06/20 09:20:45 INFO mapreduce.Job: Counters: 18
    File System Counters
        FILE: Number of bytes read=19336214
        FILE: Number of bytes written=19701015
        FILE: Number of read operations=0
        FILE: Number of large read operations=0
        FILE: Number of write operations=0
    Map-Reduce Framework
        Map input records=5
        Map output records=5
        Input split bytes=2103
        Spilled Records=0
        Failed Shuffles=0
        Merged Map outputs=0
        GC time elapsed (ms)=16
        CPU time spent (ms)=0
        Physical memory (bytes) snapshot=0
        Virtual memory (bytes) snapshot=0
        Total committed heap usage (bytes)=41136128
    File Input Format Counters
        Bytes Read=0
    File Output Format Counters
        Bytes Written=173
14/06/20 09:20:45 INFO mapreduce.ImportJobBase: Transferred 0 bytes in 28.2448 seconds
(0 bytes/sec)
14/06/20 09:20:45 INFO mapreduce.ImportJobBase: Retrieved 5 records.
```

To list the import file generated in HDFS run the following command.

```
ls -l /dump_dir
```

The output file gets listed as shown in Figure 12-22.

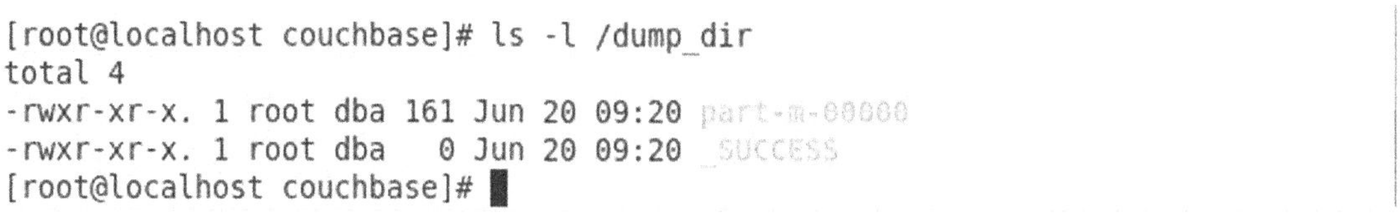

```
[root@localhost couchbase]# ls -l /dump_dir
total 4
-rwxr-xr-x. 1 root dba 161 Jun 20 09:20 part-m-00000
-rwxr-xr-x. 1 root dba   0 Jun 20 09:20 _SUCCESS
[root@localhost couchbase]# █
```

*Figure 12-22.* *Listing the files in the /dump_dir*

To list the data in the output file of the import command run the following command.

```
vi /dump_dir/part-m-00000
```

The data in the output file gets listed as shown in Figure 12-23.

```
root@localhost:/couchbase                        _ □ ✕

File  Edit  View  Search  Terminal  Help
"journal", "Oracle Magazine"
"author", "David A. Kelly"
"title", "Engineering as a Service"
"edition", "November-December 2013"
"publisher", "Oracle Publishing"
~
~
```

*Figure 12-23.* *Listing the Data in the output File*

# Importing the BACKFILL Table

The import of the BACKFILL_nn table generates an output that includes the following artifacts that are able to get streamed in the specified nn number of minutes.

- All the documents in the server.

- All mutations made in the same login session of the OS, whether while the command is running or prior to running the command.

Next, we shall import the BACKFILL_10 table, which streams key mutations for 10 minutes and also imports the key-value pairs in the Couchbase Server. Run the sqoop import command with the –table option set to BACKFILL_10.

```
sqoop import -Dmapreduce.job.max.split.locations=2048 --connect http://10.0.2.15:8091/pools
--table BACKFILL_10 --target-dir /backfill_dir --as-textfile
```

The command runs for 10 minutes as key mutations are being streamed for 10 minutes as shown in Figure 12-24.

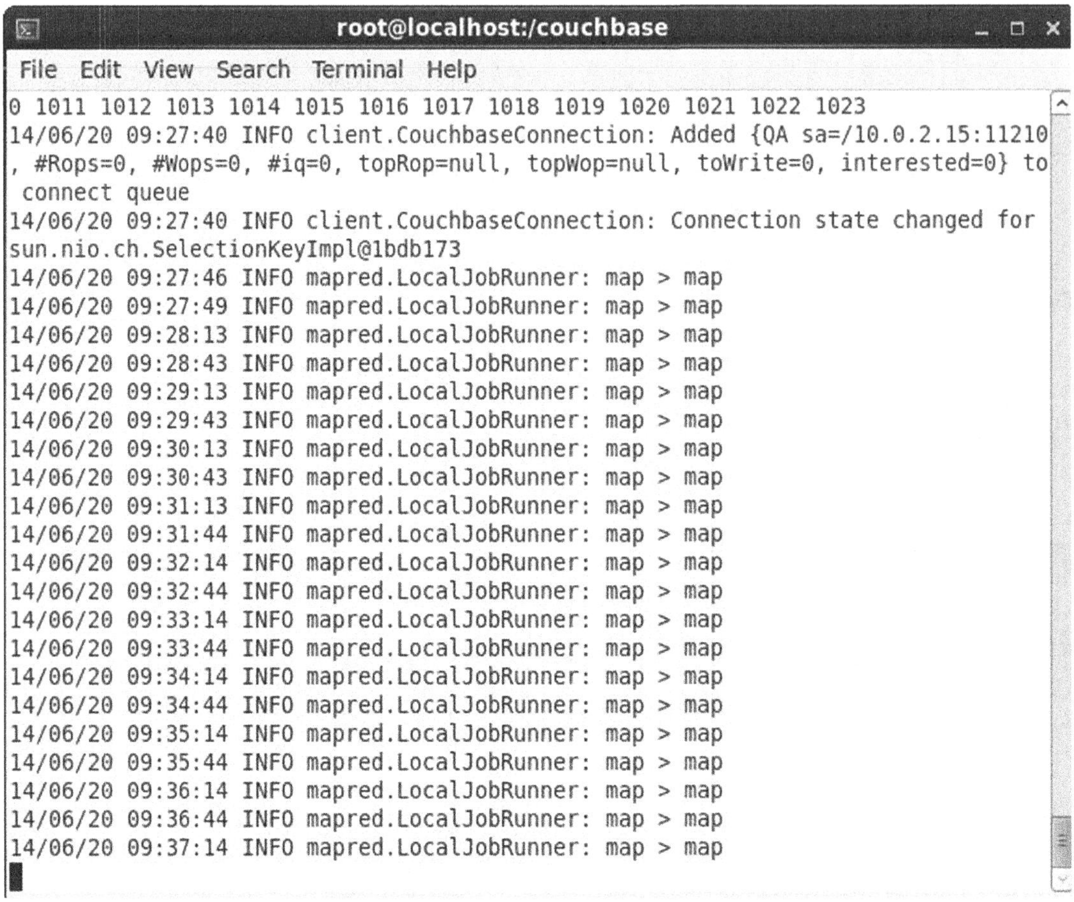

```
root@localhost:/couchbase                    _ □ ✕
 File  Edit  View  Search  Terminal  Help
0 1011 1012 1013 1014 1015 1016 1017 1018 1019 1020 1021 1022 1023
14/06/20 09:27:40 INFO client.CouchbaseConnection: Added {QA sa=/10.0.2.15:11210
, #Rops=0, #Wops=0, #iq=0, topRop=null, topWop=null, toWrite=0, interested=0} to
 connect queue
14/06/20 09:27:40 INFO client.CouchbaseConnection: Connection state changed for
sun.nio.ch.SelectionKeyImpl@1bdb173
14/06/20 09:27:46 INFO mapred.LocalJobRunner: map > map
14/06/20 09:27:49 INFO mapred.LocalJobRunner: map > map
14/06/20 09:28:13 INFO mapred.LocalJobRunner: map > map
14/06/20 09:28:43 INFO mapred.LocalJobRunner: map > map
14/06/20 09:29:13 INFO mapred.LocalJobRunner: map > map
14/06/20 09:29:43 INFO mapred.LocalJobRunner: map > map
14/06/20 09:30:13 INFO mapred.LocalJobRunner: map > map
14/06/20 09:30:43 INFO mapred.LocalJobRunner: map > map
14/06/20 09:31:13 INFO mapred.LocalJobRunner: map > map
14/06/20 09:31:44 INFO mapred.LocalJobRunner: map > map
14/06/20 09:32:14 INFO mapred.LocalJobRunner: map > map
14/06/20 09:32:44 INFO mapred.LocalJobRunner: map > map
14/06/20 09:33:14 INFO mapred.LocalJobRunner: map > map
14/06/20 09:33:44 INFO mapred.LocalJobRunner: map > map
14/06/20 09:34:14 INFO mapred.LocalJobRunner: map > map
14/06/20 09:34:44 INFO mapred.LocalJobRunner: map > map
14/06/20 09:35:14 INFO mapred.LocalJobRunner: map > map
14/06/20 09:35:44 INFO mapred.LocalJobRunner: map > map
14/06/20 09:36:14 INFO mapred.LocalJobRunner: map > map
14/06/20 09:36:44 INFO mapred.LocalJobRunner: map > map
14/06/20 09:37:14 INFO mapred.LocalJobRunner: map > map
```

*Figure 12-24.* *Importing the BACKFILL_10 Table*

If no key mutations are made in the 10 minutes, the output indicates that the ten records from the CouchbaseS have been imported to HDFS. As no key-value mutations were made in the ten minutes the command runs the output file from the command only has the key-value pairs already in the Couchbase Server and the mutations made prior to running the import command in the same login session as shown in Figure 12-25.

```
root@localhost:/couchbase                          _ □ ×
File  Edit  View  Search  Terminal  Help
cessfully
14/06/20 09:37:42 INFO mapreduce.Job: Counters: 18
        File System Counters
                FILE: Number of bytes read=19336236
                FILE: Number of bytes written=19702203
                FILE: Number of read operations=0
                FILE: Number of large read operations=0
                FILE: Number of write operations=0
        Map-Reduce Framework
                Map input records=10
                Map output records=10
                Input split bytes=2103
                Spilled Records=0
                Failed Shuffles=0
                Merged Map outputs=0
                GC time elapsed (ms)=13
                CPU time spent (ms)=0
                Physical memory (bytes) snapshot=0
                Virtual memory (bytes) snapshot=0
                Total committed heap usage (bytes)=40951808
        File Input Format Counters
                Bytes Read=0
        File Output Format Counters
                Bytes Written=334
14/06/20 09:37:42 INFO mapreduce.ImportJobBase: Transferred 0 bytes in 633.8266
seconds (0 bytes/sec)
14/06/20 09:37:42 INFO mapreduce.ImportJobBase: Retrieved 10 records.
[root@localhost couchbase]# █
```

*Figure 12-25.* *Importing the BACKFILL_10 Table with no key-value mutations*

List the files in the backfill_dir directory. The part-m-00000 is the file with the data imported from backfill_nn table as shown in Figure 12-26.

```
[root@localhost backfill_dir]# ls -l
total 4
-rwxr-xr-x. 1 root dba 322 Jun 20 09:37 part-m-00000
-rwxr-xr-x. 1 root dba   0 Jun 20 09:37 _SUCCESS
[root@localhost backfill_dir]# vi part-m-00000
[root@localhost backfill_dir]# █
```

*Figure 12-26.* *Listing the files in the backfill_dir*

Display the data in the part-m-00000 file.

```
vi /backfill_dir/part-m-00000
```

The key-value pairs in the server and the mutations made prior to running the command get displayed. As the mutations were made in the same login session the key-value pairs get listed twice as shown in Figure 12-27.

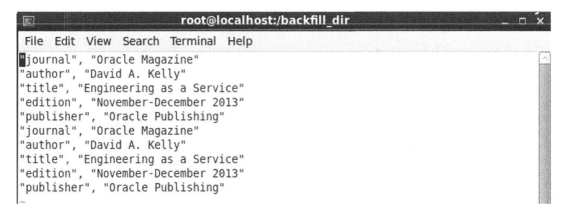

*Figure 12-27. Displaying the data in the /backfill_dir*

Next, we shall import the BACKFILL_2 table, which streams key-value mutations for 2 minutes. Delete any key-value pairs from Couchbase Server prior to running the command as we shall we demonstrating the streaming of key-value mutations while the command is running as shown in Figure 12-28.

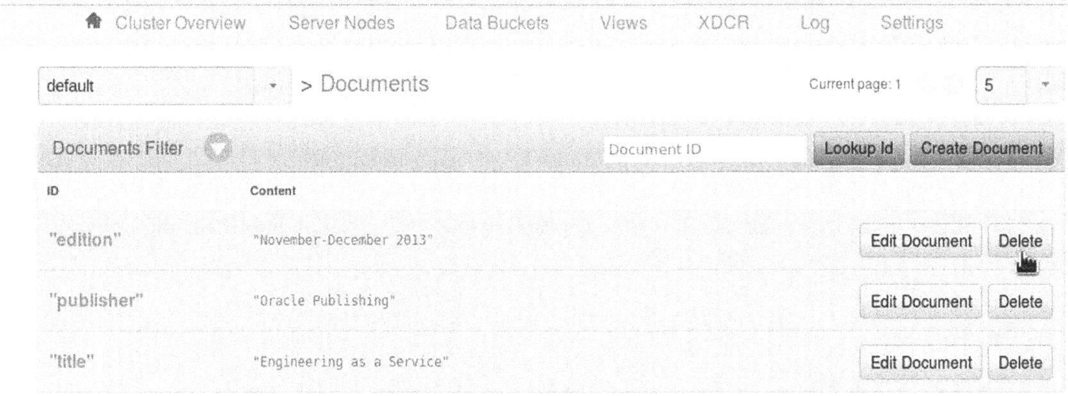

*Figure 12-28. Importing the BACKFILL_2 table*

We shall add some key-value pairs while the command is running and the key-value pairs added shall be streamed by the sqoop import command. Run the following command to stream key-value mutations for 2 minutes to HDFS.

```
sqoop import -Dmapreduce.job.max.split.locations=2048 --connect http://10.0.2.15:8091/pools
--table BACKFILL_2 --target-dir /backfill_dir_2 --as-textfile
```

The local job runner runs for two minutes streaming mutations from Couchbase Server as shown in Figure 12-29.

```
 ▣                    root@localhost:/backfill_dir                    _  □  ×

  File  Edit  View  Search  Terminal  Help

13 614 615 616 617 618 619 620 621 622 623 624 625 626 627 628 629 630 631 632 6
33 634 635 636 637 638 639 640 641 642 643 644 645 646 647 648 649 650 651 652 6
53 654 655 656 657 658 659 660 661 662 663 664 665 666 667 668 669 670 671 672 6
73 674 675 676 677 678 679 680 681 682 683 684 685 686 687 688 689 690 691 692 6
93 694 695 696 697 698 699 700 701 702 703 704 705 706 707 708 709 710 711 712 7
13 714 715 716 717 718 719 720 721 722 723 724 725 726 727 728 729 730 731 732 7
33 734 735 736 737 738 739 740 741 742 743 744 745 746 747 748 749 750 751 752 7
53 754 755 756 757 758 759 760 761 762 763 764 765 766 767 768 769 770 771 772 7
73 774 775 776 777 778 779 780 781 782 783 784 785 786 787 788 789 790 791 792 7
93 794 795 796 797 798 799 800 801 802 803 804 805 806 807 808 809 810 811 812 8
13 814 815 816 817 818 819 820 821 822 823 824 825 826 827 828 829 830 831 832 8
33 834 835 836 837 838 839 840 841 842 843 844 845 846 847 848 849 850 851 852 8
53 854 855 856 857 858 859 860 861 862 863 864 865 866 867 868 869 870 871 872 8
73 874 875 876 877 878 879 880 881 882 883 884 885 886 887 888 889 890 891 892 8
93 894 895 896 897 898 899 900 901 902 903 904 905 906 907 908 909 910 911 912 9
13 914 915 916 917 918 919 920 921 922 923 924 925 926 927 928 929 930 931 932 9
33 934 935 936 937 938 939 940 941 942 943 944 945 946 947 948 949 950 951 952 9
53 954 955 956 957 958 959 960 961 962 963 964 965 966 967 968 969 970 971 972 9
73 974 975 976 977 978 979 980 981 982 983 984 985 986 987 988 989 990 991 992 9
93 994 995 996 997 998 999 1000 1001 1002 1003 1004 1005 1006 1007 1008 1009 101
0 1011 1012 1013 1014 1015 1016 1017 1018 1019 1020 1021 1022 1023
14/06/20 09:46:35 INFO client.CouchbaseConnection: Added {QA sa=/10.0.2.15:11210
, #Rops=0, #Wops=0, #iq=0, topRop=null, topWop=null, toWrite=0, interested=0} to
 connect queue
14/06/20 09:46:35 INFO client.CouchbaseConnection: Connection state changed for
sun.nio.ch.SelectionKeyImpl@1e019cc
14/06/20 09:47:09 INFO mapred.LocalJobRunner: map > map
```

**Figure 12-29.** *Streaming mutations from Couchbase Server*

While the command is running, select Create Document in Couchbase Administration Console to create a document as shown in Figure 12-30.

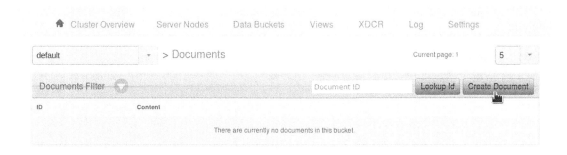

***Figure 12-30.*** *Selecting Create Document*

Specify the Document Id (`catalog1` for example) and click on Create as shown in Figure 12-31.

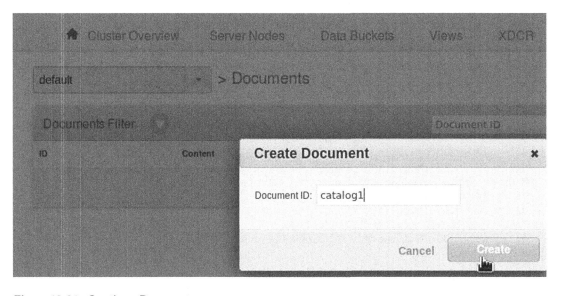

***Figure 12-31.*** *Creating a Document*

A document with Id "Catalog" gets added with the default JSON document. Similarly, add one other JSON document (`catalog2`) with non-default JSON.

```
{"journal": "Oracle Magazine",
"publisher": "Oracle Publishing",
"edition": "November-December 2013",
"title": "Engineering as a Service",
"author": "David A. Kelly"}
```

Click on Save to save `catalog2` document as shown in Figure 12-32.

***Figure 12-32.*** *Saving the catalog2 Document*

Two JSON documents `catalog1` and `catalog2` get added to Couchbase Server as shown in Figure 12-33.

***Figure 12-33.*** *Listing the JSON Documents catalog1 and catalog2*

When the 2 minutes have elapsed, the command completes and the command output indicates that six records have been imported as shown in Figure 12-34.

```
root@localhost:/backfill_dir                    _ □ ✕

File  Edit  View  Search  Terminal  Help

cessfully
14/06/20 09:54:56 INFO mapreduce.Job: Counters: 18
        File System Counters
                FILE: Number of bytes read=19336233
                FILE: Number of bytes written=19702251
                FILE: Number of read operations=0
                FILE: Number of large read operations=0
                FILE: Number of write operations=0
        Map-Reduce Framework
                Map input records=6
                Map output records=6
                Input split bytes=2103
                Spilled Records=0
                Failed Shuffles=0
                Merged Map outputs=0
                GC time elapsed (ms)=30
                CPU time spent (ms)=0
                Physical memory (bytes) snapshot=0
                Virtual memory (bytes) snapshot=0
                Total committed heap usage (bytes)=41230336
        File Input Format Counters
                Bytes Read=0
        File Output Format Counters
                Bytes Written=365
14/06/20 09:54:56 INFO mapreduce.ImportJobBase: Transferred 0 bytes in 153.0398
seconds (0 bytes/sec)
14/06/20 09:54:56 INFO mapreduce.ImportJobBase: Retrieved 6 records.
[root@localhost backfill_dir]# █
```

***Figure 12-34.*** *Importing six records from BACKFILL_2*

A more detailed output from the import command is listed:

```
14/06/20 09:52:51 INFO mapreduce.Job: The url to track the job: http://localhost:8080/
14/06/20 09:52:51 INFO mapreduce.Job: Running job: job_local2053954453_0001
14/06/20 09:52:51 INFO mapred.LocalJobRunner: OutputCommitter set in config null
14/06/20 09:52:52 INFO mapred.LocalJobRunner: OutputCommitter is org.apache.hadoop.
mapreduce.lib.output.FileOutputCommitter
14/06/20 09:52:52 INFO mapreduce.Job: Job job_local2053954453_0001 running in uber mode : false
14/06/20 09:52:52 INFO mapreduce.Job:  map 0% reduce 0%
14/06/20 09:52:52 INFO mapred.LocalJobRunner: Waiting for map tasks
14/06/20 09:52:52 INFO mapred.LocalJobRunner: Starting task: attempt_local2053954453_0001
_m_000000_0
14/06/20 09:52:54 INFO mapred.Task:  Using ResourceCalculatorProcessTree : [ ]
14/06/20 09:52:54 WARN conf.Configuration: mapreduce.map.class is deprecated. Instead, use
mapreduce.job.map.class
14/06/20 09:52:54 INFO mapred.MapTask: Processing split: 0 1 2 3 4 5 6 7 8 9 10 11 12 13 14
15 16 17 18 19 20 21 ...   1006 1007 1008 1009 1010 1011 1012 1013 1014 1015 1016 1017 1018
1019 1020 1021 1022 1023
```

14/06/20 09:52:55 INFO client.CouchbaseConnection: Added {QA sa=/10.0.2.15:11210, #Rops=0, #Wops=0, #iq=0, topRop=null, topWop=null, toWrite=0, interested=0} to connect queue
14/06/20 09:52:55 INFO client.CouchbaseConnection: Connection state changed for sun.nio.ch.SelectionKeyImpl@122d536
14/06/20 09:53:05 INFO mapred.LocalJobRunner: map > map
14/06/20 09:53:17 INFO mapred.LocalJobRunner: map > map
14/06/20 09:53:26 INFO mapred.LocalJobRunner: map > map
14/06/20 09:53:29 INFO mapred.LocalJobRunner: map > map
14/06/20 09:53:41 INFO mapred.LocalJobRunner: map > map
14/06/20 09:53:59 INFO mapred.LocalJobRunner: map > map
14/06/20 09:54:29 INFO mapred.LocalJobRunner: map > map
14/06/20 09:54:56 INFO client.CouchbaseConnection: Shut down Couchbase client
14/06/20 09:54:56 INFO db.CouchbaseRecordReader: All TAP messages have been received.

14/06/20 09:54:56 INFO mapreduce.AutoProgressMapper: Auto-progress thread is finished. keepGoing=false
14/06/20 09:54:56 INFO mapred.LocalJobRunner: map > map
14/06/20 09:54:56 INFO mapred.Task: Task:attempt_local2053954453_0001_m_000000_0 is done. And is in the process of committing
14/06/20 09:54:56 INFO mapred.LocalJobRunner: map > map
14/06/20 09:54:56 INFO mapred.Task: Task attempt_local2053954453_0001_m_000000_0 is allowed to commit now
14/06/20 09:54:56 INFO output.FileOutputCommitter: Saved output of task 'attempt_local2053954453_0001_m_000000_0' to file:/backfill_dir_2/_temporary/0/task_local2053954453_0001_m_000000
14/06/20 09:54:56 INFO mapred.LocalJobRunner: map
14/06/20 09:54:56 INFO mapred.Task: Task 'attempt_local2053954453_0001_m_000000_0' done.
14/06/20 09:54:56 INFO mapred.LocalJobRunner: Finishing task: attempt_local2053954453_0001_m_000000_0
14/06/20 09:54:56 INFO mapred.LocalJobRunner: Map task executor complete.
14/06/20 09:54:56 INFO mapreduce.Job:  map 100% reduce 0%
14/06/20 09:54:56 INFO mapreduce.Job: Job job_local2053954453_0001 completed successfully
14/06/20 09:54:56 INFO mapreduce.Job: Counters: 18
    File System Counters
        FILE: Number of bytes read=19336233
        FILE: Number of bytes written=19702251
        FILE: Number of read operations=0
        FILE: Number of large read operations=0
        FILE: Number of write operations=0
    Map-Reduce Framework
        Map input records=6
        Map output records=6
        Input split bytes=2103
        Spilled Records=0
        Failed Shuffles=0
        Merged Map outputs=0
        GC time elapsed (ms)=30
        CPU time spent (ms)=0
        Physical memory (bytes) snapshot=0
        Virtual memory (bytes) snapshot=0
        Total committed heap usage (bytes)=41230336

```
      File Input Format Counters
          Bytes Read=0
      File Output Format Counters
          Bytes Written=365
14/06/20 09:54:56 INFO mapreduce.ImportJobBase: Transferred 0 bytes in 153.0398 seconds
(0 bytes/sec)
14/06/20 09:54:56 INFO mapreduce.ImportJobBase: Retrieved 6 records.
[root@localhost backfill_dir]#
```

List the files in the backfill_dir_2 directory. The part-m-00000 file gets listed as shown in Figure 12-35.

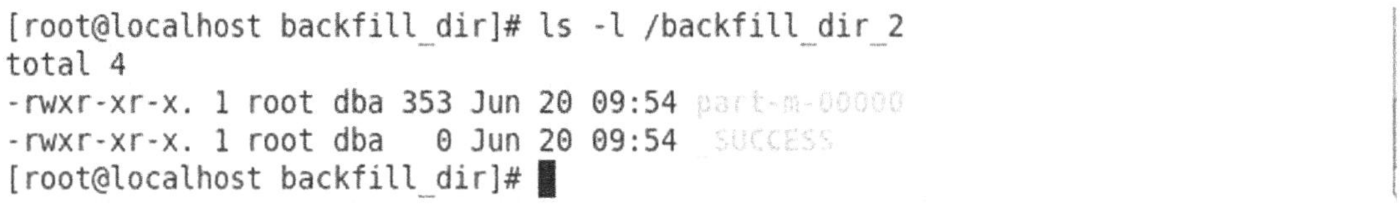

```
[root@localhost backfill_dir]# ls -l /backfill_dir_2
total 4
-rwxr-xr-x. 1 root dba 353 Jun 20 09:54 part-m-00000
-rwxr-xr-x. 1 root dba   0 Jun 20 09:54 SUCCESS
[root@localhost backfill_dir]#
```

*Figure 12-35.* *Listing files in the backfill_dir_2*

List the data in the output file generated from the sqoop import command with the following command.

```
vi /backfill_dir_2/part-m-00000
```

The data in the output file has all the key-value mutations made in the Couchbase Server while the command was running and prior to the command in the same login session. We had deleted catalog1, catalog2, and catalog3 prior to running the import command and the documents are listed in the output generated. We added catalog1 and catalog2 to the Couchbase Server while the import command was running and the documents are also streamed. We modified the catalog2 document and the modified catalog2 is listed separately in the key-value pairs streamed as shown in Figure 12-36.

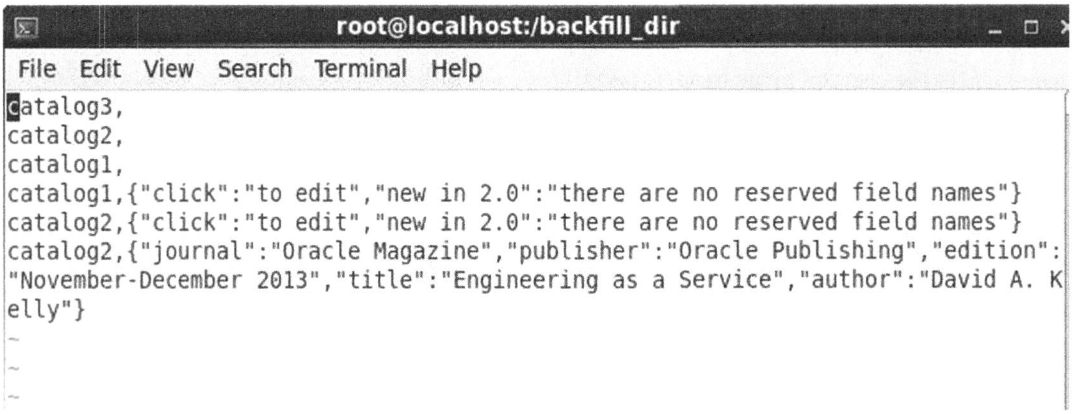

```
root@localhost:/backfill_dir

File   Edit   View   Search   Terminal   Help

catalog3,
catalog2,
catalog1,
catalog1,{"click":"to edit","new in 2.0":"there are no reserved field names"}
catalog2,{"click":"to edit","new in 2.0":"there are no reserved field names"}
catalog2,{"journal":"Oracle Magazine","publisher":"Oracle Publishing","edition":
"November-December 2013","title":"Engineering as a Service","author":"David A. K
elly"}
```

*Figure 12-36.* *Displaying the data in the backfill_dir*

# Importing JSON from Couchbase Server into HDFS

In the preceding section we imported key-value pairs in which the value was a text string. Using the BACKFILL_nn we imported key-value mutations, which included JSON. In this section we shall import key-value pairs in which the values are JSON documents. Click on Create Document in Couchbase Administration Console to create two documents "catalog1" and "catalog2" as shown in Figure 12-37. Store JSON in the two documents.

***Figure 12-37.*** *Creating two JSON Documents*

Next, import the DUMP table using the following sqoop import command.

```
sqoop import -Dmapreduce.job.max.split.locations=2048 --connect http://10.0.2.15:8091/pools
--table DUMP --target-dir /dump_dir --as-textfile
```

The output from the command indicates that two records get imported as shown in Figure 12-38.

*Figure 12-38.* *Importing the DUMP table*

Run the following command to output the data imported in the two records.

```
vi /dump_dir/part-m-00000
```

The two records imported get listed. The two records imported are the two JSON documents in the Couchbase Server as shown in Figure 12-39.

**Figure 12-39.** *Displaying the data in the /dump_dir*

# Summary

In this chapter we used the Couchbase Hadoop Connector to export data from HDFS to Couchbase Server and import data from Couchbase Server to HDFS.

This is the last chapter in the book, and you should have a learned about using Couchbase Server in web development. We discussed using Couchbase Server with commonly used languages such as Java, PHP, JavaScript, and Ruby. You learned about querying Couchbase with Elasticsearch and N1QL. You also learned about migrating MongoDB, Apache Cassandra, and Oracle Database to Couchbase Server. As Couchbase is designed to be used with big data, you learned about using Couchbase Server with the Hadoop, a big data framework.

# Index

## S, T

## U, V

## W, X, Y, Z

# Get the eBook for only $5!

Why limit yourself?

Now you can take the weightless companion with you wherever you go and access your content on your PC, phone, tablet, or reader.

Since you've purchased this print book, we're happy to offer you the eBook in all 3 formats for just $5.

Convenient and fully searchable, the PDF version enables you to easily find and copy code—or perform examples by quickly toggling between instructions and applications. The MOBI format is ideal for your Kindle, while the ePUB can be utilized on a variety of mobile devices.

To learn more, go to www.apress.com/companion or contact support@apress.com.

Printed in the United States
By Bookmasters